René R. P. de Vries, Irun R. Cohen and
Jon J. van Rood (Eds.)

The Role of Micro-organisms in Non-infectious Diseases

With 32 figures

Springer-Verlag
London Berlin Heidelberg New York
Paris Tokyo Hong Kong

René R. P. de Vries
Department of Immunohaematology and Blood Bank, University Hospital, 2333 AA Leiden, The Netherlands
Irun R. Cohen
Department of Cell Biology, The Weizmann Institute of Science, PO Box 26, 76 100 Rehovot, Israel
Jon J. van Rood
Department of Immunohaematology and Blood Bank, University Hospital, 2333 AA Leiden, The Netherlands

ISBN-13: 978-1-4471-1798-8 e-ISBN-13: 978-1-4471-1796-4
DOI: 10.1007/978-1-4471-1796-4

British Library Cataloguing in Publication data
Vries, René R. P. de
The role of micro-organisms in non-infectious diseases.
1. Man. Diseases I. Title II. Cohen, Irun R. III. Rood, Jon J. van IV. Series
616

Library of Congress Cataloging-in-Publication Data
The role of micro-organisms in non-infectious diseases/René R. P. de Vries, Irun R. Cohen, and Jon J. van Rood, eds.
p. cm. – (Argenteuil symposia) Includes index.

1. Immunopathology. 2. Communicable diseases—Complications and sequelae.
I. Vries, René R. P. de. II. Cohen, Irun R. III. Rood, Jon J. van. IV. Series.
RC582.15.R64 1990 90-9982
616'.01—dc20 CIP

Softcover reprint of the hardcover 1st edition 1990

Typeset by Wilmaset, Birkenhead, Wirral

2128/3916–543210 Printed on acid-free paper

Preface

It is becoming increasingly clear that micro-organisms are involved in the pathogenesis of many more diseases than those known today as infectious diseases. This is particularly true for autoimmune diseases and malignancies. Exciting discoveries emerge from quite different disciplines, which are not only of great scientific interest but which also open new avenues for treatment and prevention of these diseases.

Therefore we thought it was timely to convene a forum of experts from different disciplines with an emphasis on immunology and molecular biology to review and discuss "The role of micro-organisms in non-infectious dieases". Thanks to the Princess Lilian Foundation, this was realised last fall in Brussels, and this book contains all the edited papers presented at that time.

The book thus contains the state of the art of a fascinating and fast-developing field as the editors see it. We do not think that a comprehensive review would be appropriate for the subject. Instead, we tried to select a balanced set of new data and personal views of authorities from different disciplines. We are sure that any reader interested in the subject will find at least several new and intriguing facts as well as stimulating ideas in this volume.

Rêne R. P. de Vries
Irun R. Cohen
Jon J. van Rood

Contents

Contributors

J. André-Schwartz
New England Medical Center, Department of Medicine,
Hematology-Oncology Division, Boston, MA 02111, USA

K. Barrett
New England Medical Center, Department of Medicine,
Hematology-Oncology Division, Boston, MA 02111, USA

G. F. Bottazzo
Department of Immunology, The University College and Middlesex
School of Medicine, London W1P 9PG, UK

C. L. Burek
Department of Immunology and Infectious Diseases, The Johns
Hopkins School of Hygiene and Public Health, 615 N. Wolfe
Street, Baltimore, MD 21205, USA

A. Ciampolillo
Department of Immunology, The University College and Middlesex
School of Medicine, London W1P 9PG, UK

F. H. J. Claas
Department of Immunohaematology and Blood Bank, University
Hospital, 2333 AA Leiden, The Netherlands

I. R. Cohen
Department of Cell Biology, The Weizmann Institute of Science,
PO Box 2, 76 100 Rehovot, Israel

R. R. P. de Vries
Department of Immunohaematology and Blood Bank, University
Hospital, 2333 AA Leiden, The Netherlands

U. Deuss
Department of Internal Medicine II, University of Cologne, Ostmerheimerstrasse 200, 5000 Cologne 91, FRG

T. Dyrberg
Hagedorn Research Laboratory, Niels Steensens Vej 6, DK-2820 Gentofte, Denmark

M. A. Epstein
Nuffield Department of Clinical Medicine, University of Oxford, John Radcliffe Hospital, Headington, Oxford OX3 9DU, UK

R. C. Gallo
Laboratory of Tumor Cell Biology, National Cancer Institute, National Institutes of Health, Bethesda, MD 20892, USA

P. J. Heidt
Radiobiological Institute TNO, PO Box 5815, 2280 HV Rijswijk, The Netherlands

A. Herskowitz
Department of Immunology and Infectious Diseases, The Johns Hopkins School of Hygiene and Public Health, 615 N. Wolfe Street, Baltimore, MD 21205, USA

T. Koizumi
New England Medical Center, Department of Medicine, Hematology-Oncology Division, Boston, MA 02111, USA

E. H. Leiter
The Jackson Laboratory, Bar Harbor, ME 04609, USA

R. Mirakian
Department of Immunology, The University College and Middlesex School of Medicine, London W1P 9PG, UK

A. Miyazaki
Department of Immunology, The University College and Middlesex School of Medicine, London W1P 9PG, UK

N. A. Mitchison
Imperial Cancer Research Fund, Tumour Immunology Unit, University College London, Department of Biology, Medawar Building, Gower Street, London WC1E 6BT, UK

D. A. Neumann
Department of Immunology and Infectious Diseases, The Johns Hopkins School of Hygiene and Public Health, 615 N. Wolfe Street, Baltimore, MD 21205, USA

T. H. M. Ottenhoff
Department of Immunohaematology and Blood Bank, University Hospital, 2333 AA Leiden, The Netherlands

P. H. Plotz
Connective Tissue Diseases Section, Arthritis and Rheumatism Branch, National Institute of Arthritis and Musculoskeletal and Skin Diseases, National Institutes of Health

A. Puccetti
New England Medical Center, Department of Medicine, Hematology-Oncology Division, Boston, MA 02111 USA

N. R. Rose
Department of Immunology and Infectious Diseases, The Johns Hopkins School of Hygiene and Public Health, 615 N. Wolfe Street, Baltimore, MD 21205, USA

R. S. Schwartz
New England Medical Center, Department of Medicine, Hematology-Oncology Division, Boston, MA 02111 USA

G. H. Stollerman
Veterans Administration Medical Center, 200 Springs Road, Bedford, MA 01730, USA

D. W. van Bekkum
Radiobiological Institute TNO, PO Box 5815, 2280 HV Rijswijk, The Netherlands

A. J. van der Eb
Sylvius Laboratory, PO Box 9503, 2300 RA Leiden, The Netherlands

J. J. van Rood
Department of Immunohaematology and Blood Bank, University Hospital, 2333 AA Leiden, The Netherlands

W. C. A. van Schooten
Department of Immunohaematology and Blood Bank, University Hospital, 2333 AA Leiden, The Netherlands

Introduction: HLA, Plagues and People

J. J. van Rood, F. H. J. Claas and R. R. P. de Vries

In 1967 in Torino, 22 years ago, Amiel presented evidence that genes in HLA (B5 and B35) might predispose to Hodgkin's disease [1]. His study was triggered by the experiments of Lilly and co-workers who had shown in 1964 that Gross leukaemia virus induced haematological malignancy was associated with H-2 in the sense that animals which carried the H-2b haplotype were resistant, and those with the H-2k haplotype were susceptible [2]. These were the first studies linking the major histocompatibility complex (MHC) with susceptibility to micro-organism induced non-infectious disease.

The HLA community was slow in following up the lead given by Amiel, no doubt in part because his performance in the "wet" workshop preceding the conference had been less than optimal. This was the more unfair because he had very poor eyesight and his technician, who had done the HLA typing, had fallen sick. Amiel's data were later confirmed, but the HLA community only really became interested when in 1972 Stokes and co-workers reported that adult coeliac disease was associated with HLA-B8 [3] and Brewerton and co-workers [4] and Schlosstein and co-workers [5] showed in 1973 that virtually all patients suffering from ankylosing spondylitis were HLA-B27 positive. Most HLA disease associations are, however, with class II antigens, coeliac disease being the first to be so identified [6].

In the meantime studies by Benacerraf and co-workers had provided the evidence that antibody formation against a synthetic polymer was linked to a single gene [7], while McDevitt and Sela showed this to be closely linked to the H-2 complex [8,9].

On retrospect, it was only logical to extrapolate these findings to infectious diseases and to postulate that the immune response to micro-organisms would be co-determined by the MHC as well. One of the first large studies highlighting the possible importance of the HLA system for the survival of species was presented by Ceppellini at the meeting in 1972 on histocompatibility testing. The findings indicated that genes in HLA might play an important role in susceptibility for and resistance against endemic malaria [10]. Four villages in Sardinia had been studied; two in a malaria-infested region at sea-level and two in a malaria-free area in the mountains. Significant differences were found not only as expected for thalassemia and G6PD but also for HLA.

It was during these exciting years that Zinkernagel and Doherty discovered MHC restriction while studying virus-induced lymphocytic choriomeningitis in mice [11]. This provided a firm experimental basis for studies concerning the interaction of micro-organism with the MHC and the importance of this interaction for the occurrence of disease. These studies gave further support to the assumption that the MHC played a central role in the regulation of the immune response and that the extreme polymorphism of the MHC might be of pivotal importance for the adaptation of the species not only to endemic infection but also its survival during epidemics.

Support for the latter was found when a group of farmers who emigrated from The Netherlands to Surinam in the middle of the last century was studied [12,13]. The farmers and their families originated from the then only 30-year-old Netherlands. As these farmers came from between the different provinces of The Netherlands it is highly unlikely that one has to reckon with a founder effect. In other words, if we found that certain alleles occurred in the emigrants significantly more or less frequently than in the present day Dutch, it is next to impossible that this could be due to a selection by chance of emigrants having an excess or a decrease of certain alleles.

Shortly after their arrival half of the emigrants died during an epidemic of typhoid fever; 6 years later 20% of the survivors of the first epidemic died during an epidemic of yellow fever. Although no microbial (or viral) diagnostics were possible at that time, the diagnosis appears to be established beyond doubt because of the clinical course of the diseases and the post-mortem findings. After the epidemics the survivors married between themselves and were quite (re)productive. The emigrants belonged almost exclusively to a very strict Protestant church. This is probably the most important reason that only a few inter-racial marriages with the autochthonous population took place.

It was possible to identify 64 descendants of as many different families. They collaborated by giving a blood sample. Twenty-four polymorphisms were studied. As a control group 148 healthy unrelated present day Dutch were used. The majority of the controls originated from Leiden or its surrounding villages but one-third had in the last one or two generations moved from other parts of The Netherlands. Both the descendants of the surviving emigrants and the present day Dutch population were in Hardy Weinberg equilibrium. Significant differences in gene frequencies between the survivors of the epidemics and controls were found for C3, Gm, GLO and HLA-A, -B, -C and -DR. (Table 1)

Whereas the increased frequency of an allele in the descendants of the emigrants would argue for an increased resistance towards typhoid fever and/or yellow fever, the complete lack of DRw2 which is present in 29% of the present day Dutch would indicate association of this DR allele with an increased mortality. It is of special interest that the frequency of A3 was not significantly decreased, while the gene frequency of B7 was strongly decreased (from 0.16 to 0.02) but less strongly than DR2. This implicates DR2 as primarily associated with a low or at least ineffective response against *Salmonella typhi* and/or the yellow fever virus. This effect must have been dominant which is in accord with

Table 1. Significantly different gene frequencies of survivors from epidemics compared with those of a Dutch control population

Polymorphism	Alleles	Gene frequency	
		Survivors	Controls
GLO	1	0.29	0.45
	2	0.71	0.55
C3	F	0.39	0.21
	S	0.61	0.78
Gm	f(n)b	0.88	0.70
	za(x)g	0.11	0.29
HLA-A	Aw30	0.13	0.02
HLA-C	Cw3	0.26	0.17
HLA-B	B7	0.02	0.16
	B12	0.04	0.13
	B13	0.09	0.02
	Bw38	0.11	0.01
	B17	0.12	0.04
	Bw21	0.07	0.01
HLA-DR	DR2	0.00	0.16
	DR4	0.28	0.15
	DR13	0.27	0.13

the findings of others [14]. HLA-DR4 and DRw13 were primarily increased among the survivors and might thus code for a high or at least effective immunity. In other words, both susceptibility and resistance seem to be controlled by (different) dominant HLA-linked genes. Another conclusion that can be drawn from this study is that, as in the study by Ceppellini, not only for HLA but also for non-HLA polymorphisms such as GLO, C3 and Gm, differences were found which remain significant after correction for the number of comparisons made. Thus: not only HLA but also other genes influence resistance or susceptibility to microbial infections.

This book will concern not so much matters of life or death in epidemics and the role of HLA in it, but the grey area which lies between them and which we have dubbed: "the role of micro-organisms in non-infectious diseases". This grey area thus denotes those patients who are too sick to be called healthy, but not sick enough to die (yet). The MHC undoubtedly plays a role here also and many of the findings made in infectious disease should apply. It is against this historical background that we will discuss the findings of today.

In the last decade molecular biology has made its impact and has created molecular immunology. We want to be precise, characterize the cells we are working with and define the molecules which trigger their response. In doing so it seems wise not to forget that, as indicated above, there are quite a few topics which we have analysed, and thus understand, only incompletely.

One such example is the question whether it is the class I or II gene or the whole haplotype which codes for disease. It might seem naive to question the supreme importance of the gene and its product. The fact remains that most of the associations are incomplete and very dependent on the racial population studied. According to some a more consistent analysis can be performed when an (ancestral) haplotype is taken as the basic unit rather than an HLA gene alone

[15]. Epistatic effects between class II and the so-called class III genes have certainly not been ruled out definitively. All this is further complicated by the fact that some of the haplotypes are so rare that it is very difficult to imagine how they could carry a survival advantage.

Another area in which our insight is incomplete concerns the T cell repertoire. The MHC undoubtedly plays a major but not exclusive role in dictating its final profile. We have studied the question whether something like acquired neonatal tolerance as described by Billingham, Brent and Medawar in the mouse might exist in man as well. We were indeed able to show that about half of the highly immunized patients (panel reactivity against HLA 85% and over) waiting for a renal transplant had nevertheless not made antibodies against their mothers [16]. This could be due to a (neonatally) induced tolerance against noninherited maternal HLA antigens (NIMA). Next, limiting dilution experiments were performed. The number of CTL precursor cells of the 32 children from 9 families against the lymphocytes of the mother and against the lymphocytes of the father were determined. Only the NIMA and the noninherited paternal antigens (NIPA) can thus be the stimulator/target. In half of the children there is no significant difference in the CTL precursor frequency against the NIMA as compared to the NIPA. In two there were more CTL precursors against the mother than against the father. However, in half of the children of these nine families a significant lower reactivity against the maternal than against the paternal noninherited antigen was found and seven children had no detectable CTL precursors against the mother at all [17]. These data suggest that the T cell repertoire is not only governed by the self MHC antigens but also by NIMA.

The explanation for our findings could be that before, during or after delivery, maternal cells enter the bloodstream of the child and induce (partial) tolerance. It has been shown that this can change the immune response (Ir) gene profile in the mouse. It seems a priori likely that this could also happen in man. Indeed, we have preliminary data that suggest, for instance, that the occurrence (and severity) of rheumatoid arthritis can be influenced by the noninherited maternal class II antigens.

When we discuss the genetic and functional aspects of the role of microorganisms in non-infectious disease, especially on the molecular level, it appears prudent not to forget that there are many items on the population and individual level that we have analysed and understand only incompletely. Prudence is especially indicated because many of the molecular studies are so exacting in time and costs that they can only be performed in or on one or a few individuals. Especially if in such a situation observations are made which seem at variance with those made on the population level, it seems appropriate to remember that we can clone cells, but (fortunately) not individuals.

References

1. Amiel JL (1967) Study of the leucocyte phenotype in Hodgkin's disease. In: Curtoni ES, Mattiuz PL, Tosi RM (eds) Histocompatibility testing 1967. Williams and Wilkins Co, Baltimore, p 79
2. Lilly F, Boyse EA, Old LJ (1964) Genetic basis of susceptibility to viral leukaemogenesis. Lancet ii:1207

3. Stokes PL, Asquith P, Holmes GKT, Mackintosh P, Cooke WT (1972) Histocompatibility antigens associated with adult coeliac disease. Lancet ii:162
4. Brewerton DA, Caffrey M, Hart FD, James DCO, Nicholls A, Sherrock RD (1973) Ankylosing spondylitis and HL-A27. Lancet i:904
5. Schlosstein L, Terasaki PL, Bluestone R, Pearson CM (1973) High association of an HL-A antigen, w27, with ankylosing spondylitis. N Engl J Med 288:704
6. Keuning JJ, Pena AS, van Leeuwen A, van Hooff JP, van Rood JJ (1976) HLA-Dw3 association with coeliac disease. Lancet i:506
7. Kantor FS, Ojeda A, Benacerraf B (1963) Studies on artificial antigens. I. Antigenicity of DNP-polylysine and DNP copolymer of lysine and glutamic acid in guinea pigs. J Exp Med 55:55
8. McDevitt HO, Sela M (1965) Genetic control of the antibody response. I. Demonstration of determinant-specific differences in response to synthetic polypeptide antigens in two strains of inbred mice. J Exp Med 122:517
9. McDevitt HO, Tyan ML (1968) Genetic control of the antibody response in inbred mice. Transfer of response by spleen cells and linkage to the major histocompatibility locus (H-2). J Exp Med 135:1259
10. Piazza A, Belvedere MC, Bernoco D, Conighi C, Contu L, Curtoni ES, Mattiuz PL, Mayr W, Richiardi P, Scudeller G, Ceppellini R (1973) HL-A variation in four Sardinian villages under differential selective pressure by malaria. In: Dausset J, Colombani J (eds) Histocompatibility testing 1972, Munksgaard, Copenhagen, p 73
11. Zinkernagel RM, Doherty PC (1974) Restriction of in vitro T cell mediated cytotoxicity in lymphocytic choriomeningitis within a syngeneic or allogeneic system. Nature 248:701
12. de Vries RRP, Meera Khan P, Bernini LF, van Loghem E, van Rood JJ (1979) Genetic control of survival in epidemics? J Immunogenet 6:271
13. de Vries RRP, Schreuder GMTh, Naipal A, D'Amaro J, van Rood JJ (1989) Selection of typhoid and yellow fever epidemics witnessed by the DR locus. In: Dupont B (ed) Immunobiology of HLA, vol. 2: Immunogenetics and histocompatibility, Springer, New York, p 461
14. Sasazuki T, Nishimura Y, Muto M, Ohta N (1983) HLA-linked genes controlling immune response and disease susceptibility. Immunol Rev 70:51
15. Tokunaga K, Saueracker G, Kay PH, Christiansen FT, Anaud R, Dawkins RL (1988) Extensive deletions and insertions in different MHC supratypes detected by pulsed field gel electrophoresis. J Exp Med 168:933
16. Claas FHJ, Gijbels Y, Velden-de Munck Jvd, van Rood JJ (1988) Induction of B cell unresponsiveness to noninherited maternal HLA antigens during fetal life. Science 241:1815
17. van Rood JJ, Zhang L, van Leeuwen A, Claas FHJ (1989) Neonatal tolerance revisited. Immunol Lett 21:51

Section I

AUTOIMMUNE AND IMMUNOPATHOLOGICAL DISEASES

Chapter 1

Streptococci and Rheumatic Heart Disease

G. H. Stollerman

Introduction

The selection of rheumatic fever as the introductory topic for this volume may or may not be appropriate. On the one hand, few diseases have done more to stimulate the concept of "molecular mimicry" in autoimmunity. On the other hand, autoimmunity remains so far but a popular concept for the pathogenesis of this disease and is not the proven pathogenetic mechanism. The methodology of immunology has become increasingly powerful, however, and it may now be possible to identify on a molecular basis which (if any) of the group A streptococcal epitopes that are cross-reactive with human tissues are, or are not, relevant to the cardiac injury of the disease. It is this cardiac injury that makes rheumatic fever so important to the human race, and particularly to the medically neglected communities of the world [1].

I shall attempt to focus here on the nature of the infecting strain of *S. pyogenes* that causes rheumatic fever, the antigens that cross-react with host tissues, and the host factors that may determine susceptibility to the disease. In doing so, a hypothesis may be developed by which immunological advances in the study of autoimmunity may be brought to bear on rheumatic fever.

Antigens of the Streptococcal Cell

Before embarking on a review of the properties of rheumatogenic strains of group A streptococci, a schematic diagram of the cell wall of such strains might help to orient this discussion (Fig. 1.1).

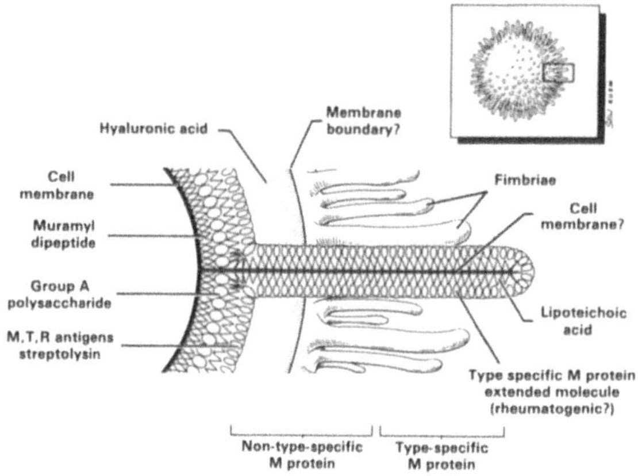

Fig. 1.1. Schematic diagram of the major known antigenic determinants on the surface of virulent, encapsulated group A streptococci.

The Cell Membrane

The cell membrane that has been studied in cell wall deficient protoplasts has not been adequately studied with regard to its anatomical relationship to surface structures such as fimbriae and its association with surface M proteins. Its cross-reactivity with heart tissues [2] will be addressed later.

Muramyl Peptide Peptidoglycans

The muramyl peptide peptidoglycans are basic cell wall building blocks. They are powerful immunological adjuvants and advantage has been taken of this property in the development of vaccines, by linking protective peptides to them. Serum antibodies to them have been found in most patients with acute rheumatic fever (ARF) [3], showing that they are themselves an antigen.

Group A Carbohydrate

Group A carbohydrate was discovered by Rebecca Lancefield and used by her to group haemolytic streptococci serologically. This carbohydrate is a weak antigen but with sensitive methods low affinity antibodies to it can be detected in ARF patients. These antibodies are particularly persistent in those with valvular heart disease [4].

Lipoteichoic Acids

Lipoteichoic acids (LTA) were once thought to be deep in the cell, and cell-membrane-associated. They are now known to be on the surface of fimbriae as well and they are the adherence factors (adhesins) of streptococci that attach to receptors consisting of fibronectin. LTA complexes electrostatically with the coiled molecules of M protein on fimbriae to form a complicated streptococcal surface that adheres to mucosal tissues and yet resists phagocytosis [5].

M Protein

M protein will be discussed in detail below as an antigen responsible for differentiating serotypes of group A streptococci, for conferring resistance to phagocytosis, and for inducing type specific immunity in the host. Its relation to autoimmunity in heart tissues and to vaccines make it of major importance. It is present in great abundance as an enlarged molecule in rheumatogenic streptococci.

Hyaluronate Capsule

The hyaluronate capsule is a feature of strains producing mucoid colonies on blood agar and is associated with highly virulent and rheumatogenic group A streptococci [1] [Fig. 1.2.). Its homology with human hyaluronate is complete, yet under certain conditions it may become antigenic and could thus be an autoantigen, of importance in the pathogenesis of ARF.

Extracellular biologically active antigens of group A streptococci are secreted into the environment of the organism in abundance. They include the following:

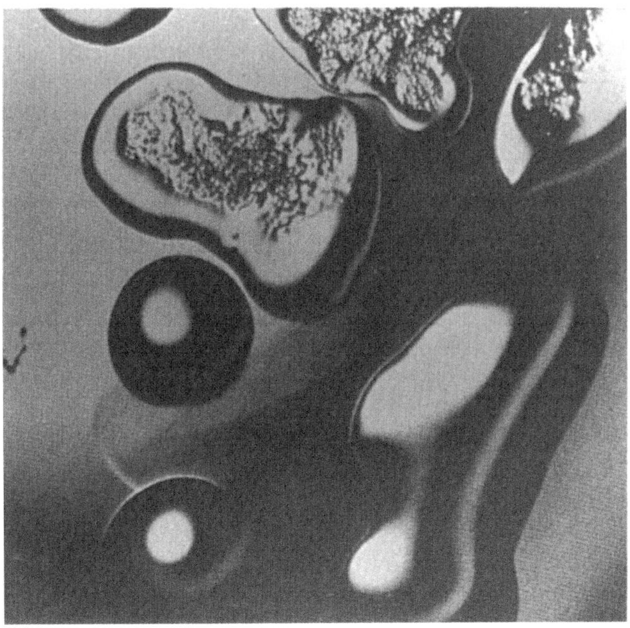

Fig. 1.2. Mucoid colonies of group A streptococci with smooth surfaces except when roughened by dehydration to a "matte" appearance. (From Wilson, A. T., *Journal of Experimental Medicine*, 109:257, 1959.)

streptolysin O, streptolysin S, hyaluronidase, streptokinase, DNases (especially B), NADases, erythrogenic toxins A, B and C, proteases, and others. They may produce pathological reactions as immune complexes, or have direct toxic effects on cells, as do the pyrogenic erythrotoxins and the streptolysins. They have not been prominent in the studies of autoimmunity.

Rheumatogenic Streptococci

About a century ago streptococci were first isolated from septic patients [1]. At about the same time rheumatic fever was perceived epidemiologically to be associated with infectious agents and particularly with epidemics of scarlet fever, whose cause at that time was still unknown. Within the next 50 years all of the clinical manifestations of rheumatic fever were finally perceived to be part of a single syndrome but it required the first 50 years of the next century to associate rheumatic fever clearly with a single species, *S. pyogenes*, among the 39 recognized species of streptococcus. The introduction of penicillin therapy and the exquisite sensitivity of *S. pyogenes* to it made the eradication of rheumatic fever by treatment and prevention of streptococcal sore throat a disarmingly simply scenario. Indeed, the decline of rheumatic fever in the developed countries of the world soon became spectacular.

To the careful student of group A streptococcal diseases, however, the above scenario was too simplistic [1]. It was noted at least 50 years ago, for example, that not all strains of group A streptococci causing sore throat, cause rheumatic fever [6]. Furthermore, certain pharyngeal M serotypes, such as M 4, failed to reactivate rheumatic fever even in highly susceptible children recently convalescent from the disease. [7]. In addition, it has long been noted that rheumatic fever and acute glomerulonephritis (AGN) are rarely associated with each other in epidemics that are clearly caused by a single strain of a known M serotype [8]. The epidemiologic dissociation of ARF and AGN was well demonstrated finally in populations in the southern United States and on the island of Trinidad, where recurrent cycles of AGN were observed without any change in the constant annual rate of rheumatic fever. Streptococcal strains causing pyoderma/impetigo were eventually differentiated clearly from those causing primarily sore throat and ARF [9]. Some strains among these pyoderma types were shown to be nephritogenic. None, however, has ever been associated with rheumatic fever [10]. M protein serotypes have now been clearly divided into "throat types" and "skin types" [11], although it is also clear that nephritogenicity and rheumatogenicity are not defined by M serotype alone [8,10] (Fig. 1.3).

Rheumatogenic Characteristics

What, then, are the characteristics of rheumatogenic streptococcal infections? First, it is clear that they must be pharyngeal [12]. No other route of infection either initiates or reactivates rheumatic fever. Secondly, it appears that intensity or virulence of infection affects the attack rate of rheumatic fever [13]. The highest attack rates recorded have been in military recruit populations in which a single virulent M serotype swept through a closed population. Within such

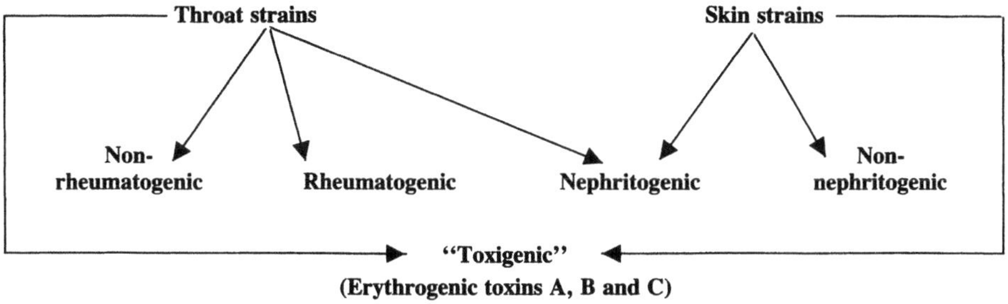

Fig. 1.3. Clinical classification of group A streptococci.

epidemics, however, it was possible to stratify the attack rate by the magnitude of the immune response in infected individuals [14]. It would appear that the intensity of the antigenic stimulus is one important determinant of the attack rate of rheumatic fever. Studies of non-epidemic, sporadic streptococcal disease in school children in the United States during the era of the striking decline in the incidence of rheumatic fever have shown that rheumatic fever disappeared despite a continued high frequency of streptococcal pharyngitis in such populations [15,16]. The absence of rheumatic fever despite this high prevalence of streptococcal sore throat in these populations cannot be explained away on the basis of treatment with penicillin alone. Almost a third of patients with rheumatic fever do not recall an antecedent sore throat and therefore would not seek medical therapy. The virulence characteristics of streptococcal strains that once caused fierce rheumatic fever epidemics, however, have certainly changed, as has been emphasized repeatedly during the past 20 years [1,10]. Several of these virulence characteristics in rheumatogenic strains have been notable: (a) They belong only to certain M serotypes. A review of the available literature has showed specific outbreaks of acute rheumatic fever (ARF) to be associated most commonly with the following M serotypes: 1, 3, 5, 6, 14, 18, 19, 24, 27 and 29 [17]. (b) Recent studies of the M protein molecule provide further insight into the relationship of M type to rheumatogenicity. M proteins of rheumatogenic and non-rheumatogenic streptococci show distinct structural differences. M protein molecules of group A streptococci that are strongly associated epidemiologically with ARF share a particularly long terminal antigenic domain [18]. (c) M proteins of the highly rheumatogenic group A streptococci so far studied contain epitopes that are shared with human heart tissue, notably sarcolemmal membrane proteins [2] and cardiac myosin [19,20]. (d) Epidemic rheumatogenic strains have been heavily encapsulated, forming large mucoid colonies on blood agar cultures [1,10]. (e) These M-rich, mucoid strains are most resistant to phagocytosis, are most virulent and are strongly immunogenic, particularly in the production of anti-M antibodies. (f) Rheumatogenic strains do not produce the opacity factor, a lipoproteinase found in many skin types [21].

The Resurgence of ARF in the United States

In the past several years, the remarkable and unanticipated resurgence of ARF in the United States has greatly strengthened the concept of specific rheumatogenic strains of group A streptococci. In early 1985, epidemic ARF was reported from Salt Lake City, Utah, and the surrounding Rocky Mountain area [22,23].

By June 74 cases, an eight-fold increase in annual incidence of ARF, were reported and all had the classic manifestations of ARF including a high frequency of carditis and Sydenham's chorea (Table 1.1). Smaller clusters of rheumatic fever ranging from 17 to 40 cases were reported during the same period from Columbus [24] and Akron, Ohio, and Pittsburgh, Pennsylvania [25]. Clinicians in many other parts of the country reported occasional cases of ARF which had not been seen on their hospital wards for many years. A survey by the Centers for Disease Control [23] identified a two-fold or greater increase in ARF in 6 of 24 states. Furthermore, the disease was not reported from urban poverty but from predominantly white middle-class children living in suburban or rural settings. Moreover, two military epidemics of ARF were reported during this period. Between December 1986 and July 1987, the Naval Training Center in San Diego, California, which had abandoned benzathine penicillin G mass prophylaxis in 1980, experienced an epidemic of streptococcal pharyngitis associated with 10 cases of ARF [26]. The second outbreak occurred at Fort Leonard Wood Army Training Base, Missouri [27]. Ten soldiers developed ARF at this base and four additional recruits developed the disease within five weeks of transfer to other Army posts. The epidemic was terminated after reinstitution of prophylaxis with intramuscular benzathine penicillin G (Fig. 1.4).

Table 1.1. Ninety-nine cases of acute rheumatic fever, by three major manifestions of Jones Criteria, Utah, 1985–1986

Major manifestations[a]	Number	Per cent
Carditis	14	14
Polyarthritis	14	14
Chorea	4	4
Carditis and polyarthritis	43	44
Carditis and chorea	14	14
Carditis, chorea, and polyarthritis	6	6
Polyarthritis and chorea	4	4
Total	99	100

[a]Categories are mutually exclusive.
From *Morbidity and Mortality Weekly Report* 36:108, 1987.

The streptococcal strains isolated from ARF patients and their contacts during these outbreaks have been those of M types well known to be associated with ARF, types 1, 3, 5, 6 and 18. Moreover, they have been mucoid strains, highly encapsulated [28], as described above for rheumatogenic M types. Of particular interest was the prominence of M type 18, a serotype rarely isolated in recent years but one dominant in military outbreaks in the past, particularly in Colorado in the 1960s. In addition to being found in ARF patients and their family contacts in various locales, M 18 was found to be prevalent in the Utah and Columbus epidemics [29]. Finally, of 85 group A streptococcal strains isolated from soldiers

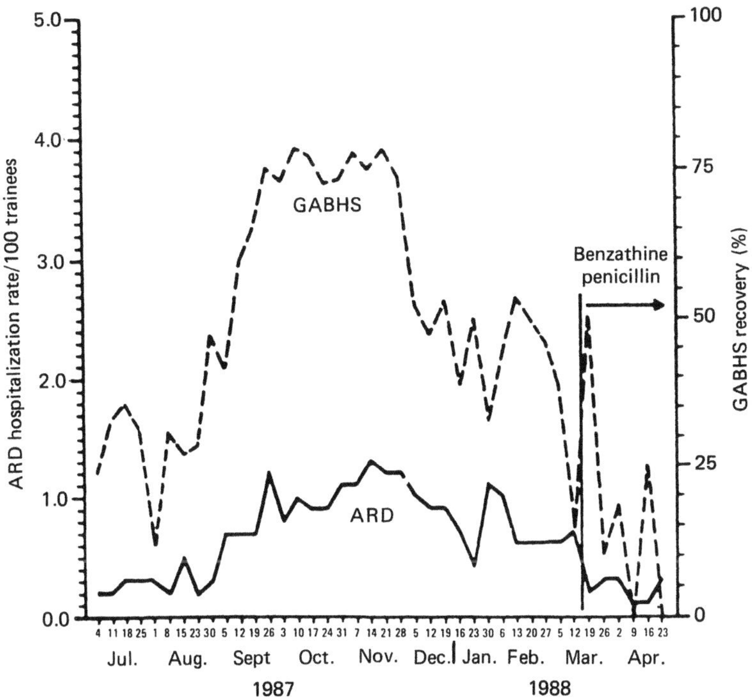

Fig. 1.4. Fort Leonard Wood, Missouri, 1987 and 1988; epidemic of streptococcal pharyngitis associated with an outbreak of rheumatic fever and terminated by mass prophylaxis with intramuscular benzathine penicillin G. (From *Morbidity and Mortality Weekly Report*, 37:519, 1988.)

from Fort Leonard Wood, 74% were mucoid M type 18. It should also be noted that unusually severe pyogenic infections, such as streptococcal pneumonia and septic arthritis appeared in these military epidemics.

In recent years, several reports of streptococcal infections producing the toxic shock syndrome have appeared [30]. Many of the strains associated with the "toxic strep syndrome" produce streptococcal pyrogenic exotoxin A, a type of erythrogenic toxin which is very similar to staphylococcus entertoxin B and which has been quite rare among streptococci causing scarlet fever in recent years [31]. The latter have produced the less potent erythrogenic toxins B and C. Thus, streptococci can and do change rapidly, and strain variation can readily explain striking variations in the incidence of the many diseases associated with *S. pyogenes*, whereas much change has not been noted in the overall prevalence of positive streptococcal throat cultures in children with sore throat [16].

Of particular importance to the understanding of the pathogenesis of ARF is the question of whether rheumatogenicity depends upon virulence factors which in turn promote sufficient antigenic stimulation by common antigenic determinants of group A streptococci to cause ARF, or whether the putative antigens causing ARF are unique to rheumatogenic strains. The former possibility makes all antigens of group A streptococci potentially suspect in ARF whereas the latter might permit us to focus on those antigens that may be unique to rheumatogenic strains.

Streptococcal Antigens Cross-reactive with Human Tissues

The immune response to streptococcal infection in the rheumatic host has been studied extensively [1]. Only some of its features that bear on autoimmunity will be considered in this review. It should be appreciated that the spectrum of clinical manifestations of ARF is very broad, ranging from a brief, non-destructive polyarthritis of a few days' duration to the acute toxic cardiac dilatation of fatal rheumatic carditis. In addition, the nature of the pathological lesions range from evanescent vasohumoral responses of erythema marginatum and the granulomas producing transient, painless subcutaneous nodules, to the intractable fibrotic process of mitral and aortic stenosis. In fact, no clear anatomical lesions can be identified with certainty in Sydenham's chorea in which CNS dysfunction is evanescent, produces no inflammatory changes in the cerebrospinal fluid, and leaves no scars.

As in the case of syphilis, hepatitis B and HIV infection, a single organism may produce the gamut of immunologic pathology. The chronicity and progression of fibrosis in some cases of rheumatic heart disease (RHD), however, suggest that unlike the above infections, in which persistence of the organism can be blamed for prolonged disease, the chronic carditis, and especially the chronic valvulitis of RHD, might be more than persistence of a non-biodegradable antigen. Some progressive valvular disease may be due to an autoimmune reaction to host tissues exposed by prior inflammatory damage, as in the case of some of the glomerulopathies. Support for this concept has been presented by Ayoub et al. [32] who have demonstrated antibodies cross-reactive with valvular glycoproteins in patients sustaining surgical valvular trauma. On the other hand, the destruction of platelets in areas of the heart exposed to valvular distortion and regurgitant jets may produce extensive fibrosis by mechanical means alone due to the liberation of fibrosis-producing platelet growth factors.

Although the detection of streptococcal antigens cross-reactive with host tissues does not necessarily imply their involvement in tissue injury, they must be looked at with suspicion if for no other reason than to ensure their elimination from candidate streptococcal vaccines. The latter, as described below, now have reasonable prospects for success because (a) protective M type specific peptide fragments of defined molecular structure can be synthesized; (b) a limited number of rheumatogenic M types make a polyvalent vaccine such as pneumococcal vaccine feasible; and (c) specific cross-reacting epitopes within such preparations are identifiable and can be excluded from vaccines. A brief review of streptococcal immunological cross-reactivity with host tissues is, therefore, in order if "molecular mimicry" between parasite and host is to be considered as a reasonable hypothesis to explain some or all of the intense immunological reactions known to occur in ARF.

Autoantibodies and ARF

A profusion of streptococcal antibodies cross-react with many human tissues, including the heart, skin, brain, glomerular basement membranes, and striated and smooth muscles [33]. As might be expected, those specific for heart tissues have received the most attention. Heart-reactive antibodies were described as early as 1945 by Cavelti [34]. Kaplan's use of the fluorescine-labelled antibody method developed with Coons [35] aroused intense interest in the autoimmune pathogenesis of rheumatic fever. Kaplan's studies showed that rabbits immunized with streptococcal cell wall developed antibodies that bound to mammalian heart tissue [36,37]. These were clearly antigens specific for group A streptococci and not the heart autoantibodies that develop secondary to cardiac necrosis, such as those seen in myocardial infarction. Kaplan believed the cross-reactive antigen to be attached to but not identical with M proteins [38]. Zabriskie and Freimer brought the concept further along in the 1960s when they described a myocardial sarcolemmal cross-reaction with group A streptococcal protoplast cell membranes which they did not believe contained any remnant of type-specific M protein [2]. Beachey and Stollerman then showed in the 1970s that non-type-specific components of M protein as well as protoplast membranes absorb heart antibodies from human sera [39]. Observations during the following years were confused by many technical problems relating to impurities of antigens and to polyclonal antibodies such as those produced by cell walls, cell membranes, and crude preparation of M proteins. A major advance in the purification of M protein was achieved by Beachey and associates in 1977 [40,41] by the gentle extraction with dilute pepsin of M protein from the surface of group A streptococci without destroying the surface fimbriae. This "pep M" extract appeared to be a molecularly homogeneous moiety with repeating peptide units. Certain of these peptide fragments contained the type specific determinants in peptides as small as 14 amino acids [42]. These type-specific peptides could be synthesized, and by developing monoclonal antibodies and cloning techniques it was possible to attack the problem of cardiac cross-reactivity with defined M protein epitopes [19,43]. Meanwhile, Krisher and Cunningham [20] identified murine monoclonal antibodies derived from animals immunized with group A streptococcal membranes which cross-reacted with cardiac myosin. The confusion between M negative membranes and M protein cross-reactions was resolved when the murine monoclonal antibodies described by Cunningham and Swerlik [44] although non-M type-specific, were shown by Dale and Beachey to react with some epitopes within the pep M 5 molecule and in cloned M 6 proteins as well [19,43]. The relevance of these studies to rheumatic fever may be enhanced by preliminary reports by Cairnes of the generation of human monoclonal antibodies cross-reactive with the group A streptococcus and with the heart derived from peripheral blood and tonsillar lymphocytes of patients with ARF [45]. However, no control studies were cited. The cross-reactive determinant also appears to be related to a heavy chain of cardiac myosin and is also non-M type-specific. It would appear possible, then, to separate M protein protective components from those cross-reactive with host tissues, although these may still be very close together on the M molecule of some strains [46].

Other Antibodies of Special Interest

No doubt other heart antigens cross-react with streptococcal antigens. The tertiary structure of M protein has been demonstrated to be that of a helical coiled-coil [47] and to be similar to other coiled proteins in human tissues which include not only myosin but keratin and perhaps other helical connective tissue structures [44]. In the 1960s, Goldstein et al. described cross-reactions between glycoproteins of heart valves and the group A polysaccharide [48]. This finding proved particularly interesting when Dudding and Ayoub found group A streptococcal antibodies to group A polysaccharide to persist in patients with rheumatic valvular disease [4]. As mentioned above, however, trauma to heart valves alone seemed to boost and sustain such antibodies [32]. Progressive valvular disease, therefore, might indeed be caused by an autoimmune reaction. However, like all other streptococcal antibodies, immunological cross-reactions with host tissues have yet to be proved to be cardiotoxic. Antibrain antibodies are found in acute rheumatic fever [49], particularly in patients with chorea. These react with the caudate nucleus and the subthalamic nuclei [50]. These antibodies can be absorbed by group A cell membranes, only partially by cell wall preparations, and not at all by group A carbohydrate. A brain antigen cross-reactive with type 24 group A streptococci appears to be a glial filament or a closely related cytoplasmic constituent [49]. This observation is of interest because neurofilaments, like myosin, tropomyosin, and keratin are, like M protein, alphahelical proteins capable of coiled-coil interactions [51].

Hyaluronic acid, the major component of the group A streptococcus capsule deserves special comment. Proven to be identical chemically with human hyaluronate, this streptococcal component was initially considered non-antigenic and was assigned relatively little significance in autoimmune disease. A preliminary report [52] of the demonstration of antibodies to streptococcal hyaluronate was considered controversial because of possible contaminating antigens in the media. Recently, however, more sensitive immunological methods (enzyme-linked immunosorbent assay; ELISA) have revealed that antibodies can be raised to highly polymerized hyaluronate that has been partially denatured [53]. This finding may have significance with respect to the rheumatogenicity of mucoid strains but reports of such antibodies in rheumatic and non-rheumatic patients have not yet appeared.

Cell-Mediated Immune Responses

In fact, careful clinical studies to explore the role, if any, of autoantibodies in causing tissue damage in ARF are quite scarce. The heart-reactive antibodies demonstrated long ago by Kaplan's [37] and Zabriskie's [54] groups did not correlate with the incidence or severity of RHD in the patients studied, nor were they shown to be cytotoxic or even specific for ARF. Indeed, the absence of such correlation with tissue damage has led to an intensive search for cross-reactive antigens that are involved in cell-mediated, cytotoxic immune responses. Several studies have suggested that cell-mediated cytotoxic reactions occur between group A streptococcal antigens and mammalian host tissues [33]. One of the earliest of these demonstrated accelerated allograft rejection by guinea-pigs hyperimmunized by group A streptococci injected in Freund's adjuvant [55]. A relationship between cell-mediated responses to streptococcal cell wall antigens and attacks of ARF has been noted [56] and depression of the absolute number of T lymphocytes and concomitant increase in B lymphocytes have been documented early in the rheumatic attack [57].

Specific damage to cardiac myofibres has been shown to occur when T cells from guinea-pigs sensitized to partially purified M protein antigens were applied to guinea-pig embryonic myofibre tissue cultures [58]. In other studies, responses were induced to syngenic cardiac determinants in the BALB/c mouse immunized with cell membranes of group A streptococci, but not with membranes of group C or G organisms or with group A streptococcal cell walls [59].

A loss of suppressor T cell function such as occurs in the experimental animal may set the stage for autoimmune disease in general. The relevance of this concept to ARF is conjectural, although significant depression of both helper and suppressor T cells occurs in children with ARF [60]. These observations may simply represent epiphenomena surrounding the delivery of a large antigenic burden to the host. Furthermore, patients with RHD without evidence of acute rheumatic inflammation may have depressed lymphocyte responsiveness in vitro that is nonspecific and that can be shown with a variety of both streptococcal and nonstreptococcal antigens [61,62]. Prospective studies of the ratio of $CD4^+$ to $CD8^+$ peripheral lymphocytes reported by Cairnes [45] show a marked temporal pattern of response. CD4 lymphocytes are reduced acutely and return to normal within 2–8 weeks whereas $CD8^+$ cells are increased from weeks 2–6, occasionally in excess of CD4. During this convalescence, these CD8 cells are shown functionally to be suppressor cells and probably represent downregulation of the vigorous immune response observed in ARF. However, an interesting recent report describes aberrant increased expression of class II antigens on heart valve fibroblasts of rheumatic patients undergoing valvular surgery [63]. These fibroblasts could represent autoantigens to infiltrating T cells and may thus trigger or exacerbate an autoimmune response. Finally, the role of the M protein molecule stimulating lymphocyte cytotoxicity for myocardial cells has also been suggested from observations of epitopes of M protein that activate S-adenosylmethionine synthetase in human lymphocytes [64]. The above observations illustrate the advances in methodology that may be brought to bear on the question of whether or not autoimmunity really is the major pathogenetic mechanism of acute rheumatic fever or whether these lymphocyte reactions are set in motion by other tissue-toxic immunological responses to streptococcal products. The structure of the streptococcal cell wall and the abundance of the extracellular substances the streptococcal cell secretes provide for a wealth of immunologic speculation about pathogenesis. The lymphocyte-stimulating properties of streptococcal "toxic" substances such as M protein moieties, streptolysin S and erythrogenic toxin A may exaggerate immune responses to all streptococcal antigens including those cross-reactive with, and cytotoxic to, host tissues.

Host Factors

The gradual acquisition of susceptibility to ARF after repeated infections in schoolchildren has suggested that rheumatic hosts have acquired increased potential for a streptococcal hyperimmune response. Every streptococcal antibody response that has been studied in ARF patients regardless of the nature of the antigen, has been exaggerated compared with control populations of patients with streptococcal pharyngitis uncomplicated by ARF [1]. Both hyperimmunity from recurrent infections and the alternative, a large antigenic burden delivered by the initiating rheumatogenic infection have been considered important in producing a pathogenetic response by the rheumatic host.

The relatively low attack rate of ARF (approximately 3%) even in the most severe epidemics of streptococcal pharyngitis, such as those seen in military populations, has focused attention on the immune response genes of the host. Heredity has long been considered a probable contributor to susceptibility to ARF because of the frequent clustering of rheumatic fever in families. The factor of infection, however, is difficult to dissociate from familial clustering due to genetic factors, and studies of identical twins have shown only limited concordance – less than that seen in atopic disease, Graves' disease, or even tuberculosis or poliomyelitis! Nonetheless, genetic host factors are still worth seeking assiduously because they are bound to contribute to the intense immunological response that is such a striking feature of the disease.

Class I HLA antigens have not been associated with ARF. HLA-DR 2, 3 and 4 have been detected with increased frequency in black, white and Indian patients respectively, primarily in those who develop carditis [65]. HLA-DR1 and DRw6 were observed with increased frequency in South African blacks [65]. In contrast to the lack of a very definitive association with HLA-DR antigens, a strong relationship has been detected with a non-HLA B cell antigen designated 883 and detected in widely distributed populations from New York to Bogota, Colombia, and from New Mexico to India [67]. Studies with a series of monoclonal antibodies directed against B cells from rheumatic fever patients [68] have identified several associated B cell alloantigens, but one of these labelled D8/17, is of particular interest [69]. It is present in a high percentage of the total B cells of rheumatic fever probands, approximately 34% compared with approximately 15% of the B cells of unaffected siblings and parents. In a set of identical twins, the proband with rheumatic fever had 43% positive B cells whereas the unaffected twin had only 15%. Thus, this B cell alloantigen is not unique to rheumatic hosts but is more consistently expressed in those who have had rheumatic fever. A reasonable hypothesis might suggest that a necessary immune response host factor, present to some degree in all individuals, is more expressible in rheumatic hosts who are stimulated by specific antigens contained in rheumatogenic group A streptococci.

Summary and Conclusions

The uniqueness of the group A streptococcus in initiating a cardiodestructive disease in a limited segment of the human species, regardless of race or ethnic group, makes the quest for a unique host response to a specific streptococcal antigen an intriguing and persisting challenge for clinical investigators, particularly for those investigators interested in autoimmunity.

New methodology is making possible more incisive research approaches. The defined streptococcal antigens that turn out to be epitopes identical with host tissues, i.e. the M protein/cardiac myosin model or the hyaluronate in the capsule of mucoid rheumatogenic strains, or others not yet known, offer the opportunity for more incisive clinical investigation. The isolation and cultivation of cardiotoxic T cell clones directed against such epitopes shared by host and parasite may eventually be possible. We may then learn more about whether autoimmunity is indeed a factor in the pathogenesis of rheumatic heart disease.

References

1. Stollerman GH (1975) Rheumatic fever and streptococcal infection. Grune and Stratton, New York
2. Zabriskie JB, Freimer EH (1966) An immunological relationship between the group A streptococcus and mammalian muscle. J Exp Med 124:661–678
3. Bahr GM, Majeed HA, Yousof AM, Chedid L, Behbehami K (1986) Detection of antibodies to muramyl dipeptide (MDP), the adjuvant moiety of streptococcal cell wall, in patients with rheumatic fever. J Infect Dis 154:1012–1017
4. Dudding BA, Ayoub EM (1968) Persistence of streptococcal group A antibody in patients with rheumatic valvular disease. J Exp Med 128:1081–1098
5. Beachey EH, Ofek I (1976) Epithelial cell binding of group A streptococcal by lipoteichoic acid on fimbriae denuded of M protein. J Exp Med 143:759–771
6. Coburn AF, Pauli RH (1935) Studies on the immune response of the rheumatic subject and its relationship to activity of the rheumatic process. IV. Characteristics of strains of hemolytic streptococcus, effective and non-effective in initiating rheumatic activity. J Clin Invest 14:755–762
7. Kuttner AG, Krumwiede E (1941) Observations on the effect of streptococcal upper respiratory infections in rheumatic children. A three-year study. J Clin Invest 20:273–287
8. Stollerman GH (1969) Nephritogenic and rheumatogenic group A streptococci. J Infect Dis 120:258–263
9. Wannamaker LW (1970) Medical progress. Differences between streptococcal infections of the throat and of the skin. N Engl J Med 282:23–30, 78–85
10. Stollerman GH (1975) The relative rheumatogenicity of strains of group A streptococci. Mod Concept Cardiovasc Dis 44:35–40
11. Dillon HC (1972) Streptococcal infections of the skin and their complications: impetigo and nephritis. In: Wannamaker LW, Milton JM (eds) Streptococci and streptococcal diseases. Academic Press, New York, pp 571–587
12. Wannamaker LW (1973) The chain that links the heart to the throat. Circulation 48:9–18
13. Stollerman GH (1961) Factors determining the attack rate of rheumatic fever. J Am Med Ass 177:823–828
14. Stetson CA (1954) The relation of antibody response to rheumatic fever. In: McCarthy M (ed) Streptococcal infections. Columbia University Press, New York, pp 208–218
15. Siegel AC, Johnson EE, Stollerman GH (1961) Controlled studies of streptococcal pharyngitis in a pediatric population: I. Factors related to the attack rate of rheumatic fever. N Engl J Med 265:559–566
16. Holmberg SD, Faich GA (1983) Streptococcal pharyngitis and acute rheumatic fever in Rhode Island. J Am Med Ass 250:2307–2312
17. Bisno AL (1980) The concept of rheumatogenic and non-rheumatogenic group A streptococci. In: Read SE, Zabriskie JB (eds) Streptococcal diseases and the immune response. Academic Press, New York, pp 789–803

18. Bessen D, Jones KF, Fischetti VA (1989) Evidence for two distinct classes of streptococcal M-protein and their relationship to rheumatic fever. J Exp Med 169:269–283
19. Dale JB, Beachey EH (1986) Sequence of myosin cross-reactive epitopes of streptococcal M protein. J Exp Med 164:1785–1790
20. Krisher K, Cunningham MW (1985) Myosin: a link between streptococci and heart. Science 227:413–415
21. Widdowson JP, Maxted WR, Grant DL (1970) The production of opacity in serum by group A streptococci and its relationship with the presence of M antigen. J Gen Microbiol 61:343–353
22. Veasy LG, Wiedmeier SE, Orsmond GS et al. (1987) Resurgence of acute rheumatic fever in the intermountain area of the United States. N Engl J Med 316:421–427
23. Centers for Disease Control (1987) Acute rheumatic fever – Utah. MMWR 36(9):108
24. Congeni B, Rizzo C, Congeni J, Screenivasan VV (1987) Outbreak of acute rheumatic fever in Northeast Ohio. J Pediat 111:176–179
25. Wald ER, Dashefsky B, Feidt CL et al. (1987) Acute rheumatic fever in Western Pennsylvania and the Tri-state Area. Pediatrics 80:371–374
26. Centers for Disease Control (1988) Acute rheumatic fever at a Navy training center – San Diego, California. MMWR 37:101
27. Centers for Disease Control (1988) Acute rheumatic fever among Army trainees – Fort Leonard Wood, MO, 1987–1988. MMWR 37:519–522
28. Marcon MJ, Hribar MM, Hosier DM et al. (1988) Occurrence of mucoid M-18 *Streptococcus pyogenes* in a Central Ohio pediatric population. J Clin Microbiol 26:1539–1542
29. Kaplan EL, Johnson DR, Cleary PP (1989) Group A streptococcal serotypes isolated from patients and sibling contacts during the resurgence of rheumatic fever in the United States in the mid-1980s. J Infect Dis 159:101–103
30. Stevens DL, Tanner MH, Winship J et al. (1989) Severe group A streptococcal infections associated with a toxic shock-like syndrome and scarlet fever toxin A. N Engl J Med 321:1–7
31. Johnson LP, L'Italier JJ, Schlievert PM (1986) Streptococcal pyrogenic exotoxin type A (scarlet fever toxin) is related to *Staphylococcus aureus* entertoxin B. MGG 203:354–356
32. Ayoub EM, Taranta A, Bartley TD (1974) Effect of valvular surgery on antibody to the group A streptococcal carbohydrate. Circulation 50:144–150
33. Stollerman GH (1987) Autoimmunity and rheumatic fever. In: Cohen IR (ed) Perspectives on autoimmunity. CRC Press, Boca Raton, FL, pp 143–153
34. Cavelti PA (1945) Autoantibodies in rheumatic fever. Proc Soc Exp Biol Med 60:379–381
35. Coons AH, Kaplan MH (1950) Localization of antigen in tissue cells. II. Improvements in a method for the detection of antigen by means of fluorescent antibody. J Exp Med 91:1–13
36. Kaplan MH, Suchy ML (1964) Immunological relation of streptococcal and tissue antigens. II. Cross-reactions of antisera to mammalian heart tissue with a cell wall constituent of certain strains of group A streptococci. J Exp Med 119:643–650
37. Kaplan MH, Svec KH (1964) Immunologic relation of streptococcal antibody cross-reactive with heart tissue. Association with streptococcal infection, rheumatic fever and glomerulonephritis. J Exp Med 119:651–666
38. Kaplan MH (1969) Cross-reaction of group A streptococci and heart tissue: Varying serologic specificity of cross-reactive antisera and relation to carrier-hapten specificity. Transplant Proc 1:976–980
39. Beachey EH, Stollerman GH (1973) Mediation of cytotoxic effects of streptococcal M protein by non-type specific antibody in human sera. J Clin Invest 52:2563–2570
40. Cunningham M, Beachey EH (1975) Immunochemical properties of streptococcal M protein purified by isoelectric focusing. J Immunol 115:1002–1006
41. Beachey EH, Stollerman GH, Chiang EY, et al. (1977) Purification and properties of M protein extracted from group A streptococci with pepsin. Covalent structure of the amino terminal region of type 24 M antigen. J Exp Med 145:1469–1483
42. Beachey EH, Tartar A, Seyer JM, Chedid L (1984) Epitope-specific protective immunogenicity of chemically synthesized 13-, 18-, and 23-residue peptide fragments of streptococcal M protein. Proc Natl Acad Sci USA 81:2203–2207
43. Dale JB, Beachey EH (1986) Localization of protective epitopes of the amino acid terminus of type 5 streptococcal M protein. J Exp Med 163:1191–1202
44. Cunningham MW, Swerlick RA (1987) Poly specificity of antistreptococcal murine monoclonal antibodies and their implications in autoimmunity. J Exp Med 164:998–1012
45. Cairnes LM (1987) Immunological studies in rheumatic fever. The immunology of rheumatic fever. N Z Med J 101:388–391

46. Bronze MS, Beachey EH, Dale JB (1988) Protective and heart cross-reactive epitopes located within the NH2 terminus of type 19 streptococcal M protein. J Exp Med 167:1849–1859
47. Manjula BN, Trus BL, Fischetti VA (1985) Presence of two distinct regions in the coiled-coil structure of the streptococcal pep M5 protein relationship to mammalian coiled-coil proteins and implications to its biological properties. Proc Natl Acad Sci USA 82:1064–1068
48. Goldstein I, Halpern B, Robert L (1967) Immunological relationship between streptococcus A polysaccharide and the structural glycoproteins of heart valve. Nature 213:44–47
49. Dorling J, Kingston J (1976) Anti-streptococcal antibodies reacting with brain tissue. II. Ultrastructural studies. Br J Exp Path 57:255–265
50. Husby G, van de Rijn I, Zabriskie J, Abdin Z, Williams R Jr (1976) Antibodies reacting with cytoplasm of subthalamic and caudate nuclei neurons in chorea and acute rheumatic fever. J Exp Med 144:1094–1110
51. McKeon FD, Kirschner MW, Caput D (1986) Homologies in both primary and secondary structure between nuclear envelope and intermediate filament proteins. Nature 319:463–468
52. Sandson J, Hamerman D, Janis R, Rojkind M (1968) Immunologic chemical similarities between the streptococcus and human connective tissue. Trans Ass Am Physns 81:249–257
53. Fillit HM, Blake M, MacDonald C, McCarthy M (1988) Immunogenicity of liposome-bound hyaluronate in mice. J Exp Med 168:971–982
54. Zabriskie JB, Hsu KC, Seegal BC (1970) Heart-reactive antibody associated with rheumatic fever: characterization and diagnostic significance. Clin Exp Immunol 7:147–159
55. Chase RM, Rapaport FT (1965) The bacterial indirection of homograft sensitivity. I. Effects of sensitization with group A streptococci. J Exp Med 122:721–732
56. Read SE, Fischetti VA, Utermohlen V, Falk RE, Zabriskie JB (1974) Cellular reactivity studies to streptococcal antigens. Migration inhibition studies in patients with streptococcal infections and rheumatic fever. J Clin Invest 54:439–450
57. Leuker RD, Abdin ZH, Williams RC Jr (1975) Peripheral blood T and B lymphocytes during acute rheumatic fever. J Clin Invest 55:975–985
58. Yang LC, Soprey PR, Wittner MK, Fox EN (1977) Streptococcal-induced cell-mediated immune destruction of cardiac myofibers in vitro. J Exp Med 146:344–360
59. Senitzer D, Cafrumy W, Pansky B, Freimer EH (1977) Spontaneous and induced cell-mediated reactivity to syngenic cells. Nature 268:158–159
60. Williams RC Jr, Raizada V, Prakash K, van de Rijn I, Zabriskie JB, Stobo JD, Sharma KB (1982) Changes in T-lymphocyte subsets during acute rheumatic fever. J Clin Immunol 2:166–172
61. Gray ED, Wannamaker LW, Ayoub EM, Elkholy A, Abdin ZH (1981) Cellular immune responses to extracellular streptococcal products in rheumatic heart disease. J Clin Invest 68:665–671
62. Shiokawa Y, Iida V, Taniguchi K (1978) Cellular immunity in rheumatic heart disease. Jpn Circ J 42:39–42
63. Amoils B, Morrison RC, Wadee AA et al. (1986) Aberrant expression of *HLA-DR* antigen on valve fibroblast from patients with active rheumatic carditis. Clin Exp Immunol 66:88–94
64. Kotb M, Dale JB, Beachey EH (1987) Stimulation of S-adenosylmethionine synthetase in human lymphocytes in streptococcal M protein. J Immunol 139:202–206
65. Ayoub EM, Barrett DJ, Maclaren MC et al. (1986) Association of class II human histocompatibility leukocyte antigens with rheumatic fever. J Clin Invest 77:2019–2026
66. Maharaj B, Hammond MG, Appadoo B et al. (1987) *HLA-A*, *B*, *DR* and *DO* antigens in black patients with severe chronic rheumatic heart disease. Circulation 76:259–261
67. Patarroyo ME, Winchester RJ, Vejerano A, et al. (1979) Association of a B-cell alloantigen with susceptibility to rheumatic fever. Nature 278:173–174
68. Zabriskie JB, Lavenchy D, Williams RC Jr, Fu SM, Yeadon CA, Fontino M, Braun DG (1985) Rheumatic fever-associated B cell alloantigens as identified by monoclonal antibodies. Arthritis Rheum 28:1047–1051
69. Khanna AK, Buskirk DR, Williams RC Jr et al. (1989) Presence of a non-*HLA* B cell antigen in rheumatic fever patients and their families as defined by a monoclonal antibody. J Clin Invest 83:1710–1716

Chapter 2

Viruses and Myocarditis

N. R. Rose, D. A. Neumann, C. L. Burek and A. Herskowitz

Introduction

Autoimmune disease has often been postulated to result from the action of micro-organisms. The first autoimmune disease described in humans, paroxysmal cold haemoglobinuria, was associated with syphilis. Through the years a number of other enigmatic diseases has been cited as the result of post-infectious autoimmune responses. Examples include rheumatic fever, which is closely related to streptococcal infection, the encephalitis that occasionally follows measles, and the orchitis associated with mumps virus infection. In this chapter we will describe the evidence that some forms of myocarditis represent an autoimmune response to cardiac antigens arising as a consequence of coxsackie virus infection.

A number of steps is necessary to connect an infectious agent with a non-infectious sequela and to elucidate the pathogenetic mechanisms: (a) There should be an epidemiological association of the particular infectious agent with the disease. (b) Immunological evidence of recent or remote infection should be obtained. Generally, this evidence is obtained through demonstration of a rising titre of antibody or through demonstration of a cell-mediated response to antigen of the candidate agent. (c) The infectious form of the agent should not be present in the diseased organ. The infectious agent will be present during earlier stages of the disease and it is quite possible that the organism (especially if it is a virus) persists in a latent non-infectious form throughout the disease process. (d) Evidence of a humoral or cellular autoimmune response specific for the diseased organ should be obtained. (e) One or more antigens of the organ should be defined as the precise target of the autoimmune response. (f) It should be possible to reproduce the disease by active immunization with the appropriate antigen or antigens. Such experiments would normally be carried out in an appropriate experimental animal. A number of different animal species, or inbred strains of the same species, should be tested. This step is particularly crucial because individuals vary greatly in their ability to develop an autoimmune response. The non-infectious sequelae of infectious disease probably represent a genetically determined abnormal response which is present in only a small proportion of the total population. (g) The availability of an experimental model will permit definition of the autoantigenic determinants as well as further investigations of the immunopathic mechanisms of disease, especially the relative roles of humoral and cellular factors. (h) It should be possible to transfer the experimental disease by passive or adoptive immunization. In vitro models may be used for investigations of the effector mechanisms. (i) Finally, it should be possible to transfer much of the experimental data to the human disease.

Delineation of the responsible antigen is the first step in developing a diagnostic serological test for the autoimmune condition. Genetic studies may permit the identification of individuals or populations with great risk of developing the non-infectious complication. Most important, understanding the pathogenetic mechanisms opens opportunities for specific treatment and prevention based on abrogating or avoiding the autoimmune response.

The Role of Coxsackie Viruses in Heart Disease

The principal diseases of the heart muscle that have been associated with virus infection are acute myocarditis and dilated cardiomyopathy. Acute myocarditis is characterized by myocardial insufficiency of sudden onset but may be associated with a history of an influenza-like illness. The clinical features include peripheral muscle pains, tachycardia, gallop rhythm and raised jugular venous pressure. A pericardial friction rub may be detected and in severe cases there may be pulmonary oedema, hypotension and circulatory collapse. In such cases, a weak thready pulse, peripheral cyanosis, sweating and anxiety are seen; complete or partial heart block and various types of arrhythmias may occur. Electrocardiography reveals dilated ventricles, poor systolic function, low stroke volume, and diminished cardiac output. Death is not common, but may occur, especially if the patient undertakes strenuous exercise.

At least 90% of patients recover fully from acute myocarditis. A few die in the acute stage and some progress to dilated cardiomyopathy. The clinical features of dilated cardiomyopathy are those of congestive heart failure, i.e. dilatation of the left ventricle, gallop rhythm and impaired left ventricular function on echocardiography. The prognosis of dilated cardiomyopathy is poor, with a 5-year survival rate of about 50%. The diagnosis of myocarditis and dilated cardiomyopathy has been greatly facilitated by the introduction of endomyocardial biopsy. The principal histological features of myocarditis are mononuclear cell infiltration, myocyte necrosis, and eventually replacement fibrosis and calcification. The histological signs of dilated cardiomyopathy are less characteristic, but may comprise some evidence of previous inflammatory myocarditis together with endomyocardial fibrosis. Because of the focal characteristics of the infiltrate and necrosis, and because only a small portion of the myocardium is accessible to biopsy, it is possible to overlook many cases of myocarditis. This suggests that alternative diagnostic procedures are needed in order to define the patient population.

Many different viruses have been implicated in myocarditis and cardiomyopathy. The coxsackie viruses are the most common of these agents. The association is based on a history of respiratory or enteric symptoms as well as a rising titre of antibody to coxsackie B viruses. One well documented outbreak was described by Helin et al., in which 18 patients developed myopericarditis following coxsackie infection [1]. Five per cent of the symptomatic individuals exhibited persistent cardiac abnormalities, suggesting that only a small minority of patients with coxsackie virus infection developed symptomatic heart disease.

The cardiac manifestations of group B coxsackie virus infection are more severe in infants younger than 3 months of age, who may develop fatal myocarditis. Infection may actually occur before birth with the virus crossing the placenta from the infected mother to the fetus. Severe neonatal infection rarely lasts longer than 14 days after onset, and death, when it occurs, is usually 2–5 days after infection.

In older age groups the disease is less often fatal. The initial symptoms are usually respiratory or gastrointestinal with the first evidence of acute heart disease appearing 7–10 days later.

The connection of coxsackie virus infection with chronic cardiomyopathy is less certain. Sometimes there is a history of viral illness preceding the development of heart failure. Neutralizing antibody to coxsackie virus is demonstrable in a few patients [2]. Most investigators have been unable to isolate virus or detect viral particles following cardiac biopsy of cases of chronic dilated cardiomyopathy. There have been, however, reports of coxsackie viral RNA sequences demonstrable by molecular hybridization [3].

There has been a number of reports of cardiac autoimmunity in patients with acute myocarditis or congestive cardiomyopathy including elevated titres of anti-actin and antimyosin antibodies [4–7]. The heart-reactive antibodies could bind complement and produce lysis of rat myocytes. In addition, lymphocytes from patients underwent proliferation in response to cardiac antigens. Other investigators have demonstrated immune complexes in the interstitial spaces of the myocardium in patients with acute myocarditis [8]. Schultheiss and his colleagues found antibodies to the adenine nucleotide translocator protein which correlated with the decreased ejection fraction in patients with dilated cardiomyopathies [9]. Thus, there is ample evidence of autoimmunity to cardiac antigens in patients with myocarditis and cardiomyopathy.

In collaboration with Dr Kenneth Baughman we have had the opportunity of examining patients admitted to The Johns Hopkins Hospital with a clinical diagnosis of cardiomyopathy. The sera were examined by indirect immunofluorescence using 1 : 5, 1 : 10 and 1 : 20 dilutions of patient serum and a substrate of frozen rat cardiac muscle. Over 50% of cardiomyopathy patients had intermediate (10) or high (20) titres compared with 19% of normal control subjects (Fig. 2.1) sera were positive for heart reactive antibodies. Three different patterns of reaction could be discerned: antifibrillary, antisarcolemmal and diffuse, anti-intercalated disk. The appearance of each of these reaction patterns is shown in Figs 2.2. and 2.3. These results show that heart-reactive antibodies are somewhat more prevalent in patients with cardiomyopathy than in normal controls and suggest that more than one cardiac antigen is involved. It remains for future work to identify these antigens and determine their relevance to disease.

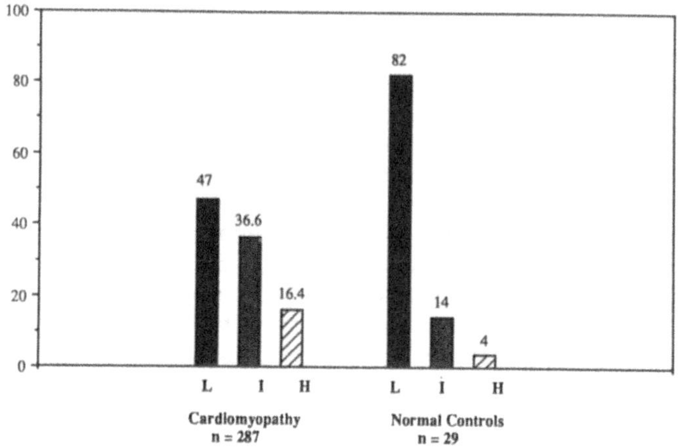

Fig. 2.1. Prevalence of heart-reactive antibodies in serum from healthy individuals (normal controls) and from cardiomyopathy patients. Serum samples were serially diluted, incubated with frozen sections of normal rat heart, and heart-reactive antibodies detected by a fluorescein-labelled second antibody specific for human IgG. Titres of <1 : 10 were considered low (*L*), those between 1 : 10 and 1 : 20 were considered intermediate (*I*), and those >1 : 20 were considered high (*H*).

Fig. 2.2. Representative fibrillary staining of normal rat heart by human serum. The fluorescein-labelled second antibody is specific for IgG.

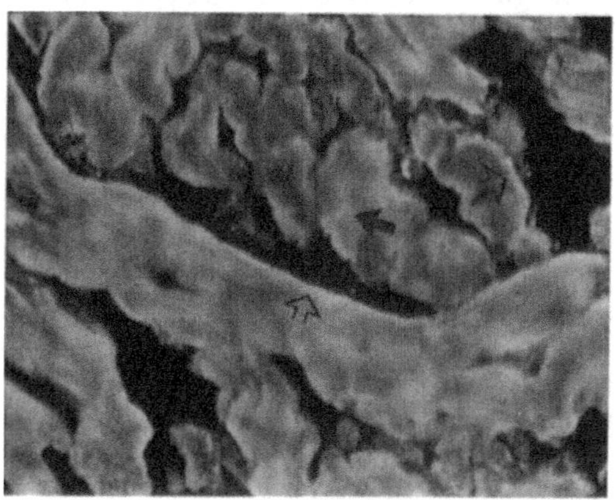

Fig. 2.3. Representative sarcolemmal (*open arrows*) and diffuse (*solid arrow*) staining of normal rat heart by human serum. The fluorescein-labelled second antibody is specific for IgG.

Coxsackie-Induced Myocarditis

Based on the pioneering studies of Woodruff [10] and of Reyes et al. [11], we adopted a murine model designed to detect the immunopathological mechanism reponsible for post-infectious myocarditis. For this purpose we infected 2-week-old mice with coxsackie virus B3 (CB3) and examined tissue samples from infected and control uninfected animals at various times after infection [12]. Tissues were studied for their content of infectious virus as well as for histological changes. Serum samples were tested for neutralizing antibody to virus and autoantibody to relevant tissues. We were able to examine large panels of 24 inbred mouse strains obtained from either Jackson Laboratory, Bar Harbor, Maine, or Dr C. S. David, Mayo Medical School, Rochester, Minnesota.

The first histological changes in the heart were seen on the third day after infection. In general, the histopathological changes were similar in all of the genetically diverse strains tested [13]. The early lesions were focal in nature and characterized by loss of myofibrillary definition, dense eosinophilic changes and disruption of the myocyte membranes. These histological changes were associated with focal cellular infiltrates predominantly of polymorphonuclear neutrophils with occasional macrophages (Fig. 2.4). By day 15 focal lesions became fibrotic with evidence of calcification (Fig. 2.5). Gradually the histological picture changed. In many mouse strains there was no further evidence of active inflammation. By day 21 the inflammatory process largely had resolved, leaving little evidence of the earlier disease.

In the course of this early phase of disease there were marked differences among the different strains of mice with respect to the level and duration of viremia, the content and persistence of virus in the heart, spleen, thymus, and pancreas, and the time of appearance of neutralizing antibodies [12]. The more susceptible strains such as A.BY and A.SW had levels of virus in their blood greater than 10^4 ml^{-1} 2 and 3 days after infection. In contrast, the more resistant strains such as A.CA, B10.S, B10.PL, and C3H and C57BL had significantly lower virus levels. Day 3 was the critical point for the appearance of neutralizing antibody since all of the resistant strains had detectable neutralizing antibody at this time, whereas the susceptible strains had no detectable neutralizing antibody to CB3 on day 3. However, the final titre of neutralizing antibody was approximately the same in all the strains examined. In general, the virus content of spleen, thymus and pancreas was greater in the susceptible strains, although the amount of CB3 recovered from the heart did not differ statistically among strains. In all strains the greatest virus content of the heart was found on day 5, the time of most severe myocardial injury. No infectious virus was detected in the heart after day 15. In contrast, infectious virus was found in the pancreas as long as 21 days after infection.

These results make it clear that there are genetic differences in the susceptibility of mice to early acute myocarditis. The susceptibility is controlled by genes both within and outside of the major histocompatibility complex (MHC) [14]. A comparison of the A congenic strains showed that while A.BY and A.SW are highly susceptible, A.CA mice are resistant to the early phase of disease. This difference in susceptibility is due to genes within the MHC, the only region where these mice differ. There are also significant differences in susceptibility between A.SW and B10.S strains which share the same H-2^s haplotype. These differences show that non-H-2 (background) genes also contribute to resistance or susceptibi-

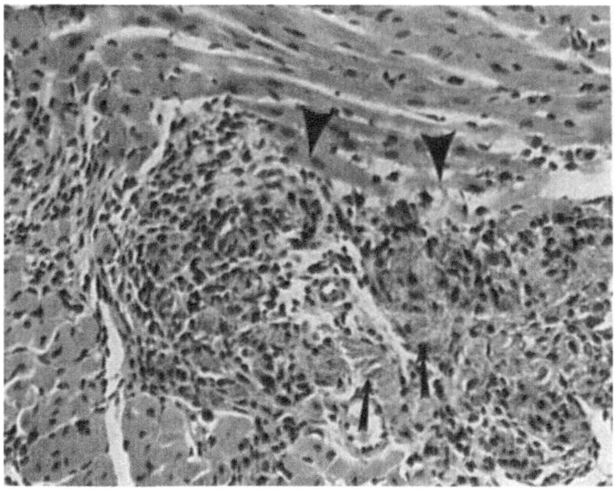

Fig. 2.4. Medium power photomicrograph taken from the right ventricular free wall of an A/J mouse 9 days after infection with CB3. Note the large focus of inflammation engulfing a cluster of necrotic myocytes. The necrotic focus typically has a sharp and distinct border with normal adjacent myocytes (*arrowheads*). In addition to the mixed inflammatory cell infiltrate, degenerated myocytes which appear granular represent early calcification (*arrows*). Haematoxylin and eosin, 440 ×.

lity to the early phase of disease. The balance between susceptibility and resistance is related to the time of appearance of neutralizing antibody.

As mentioned above, most strains of mice completely resolve their disease after day 21. However, a second distinct phase of cardiac pathology became evident in a few strains 9 days after infection and was fully manifested 15–21 days after infection. A diffuse interstitial infiltrate characterized this later stage of myocarditis comprising principally macrophages and lymphocytes (Fig. 2.6).

Fig. 2.5. Photomicrograph showing typical features of a healing focus of myocyte necrosis in a B10.A mouse 28 days after infection with CB3. Note the deposition of interstitial connective tissue, filling in areas of myocyte dropout. A sparse mononuclear infiltrate persists within the organizing scar. Note the extension of the fibrotic response to adjacent myocytes (*arrowheads*) and the dense granular staining within the lesion (*arrows*), representing more mature calcification. 440 ×.

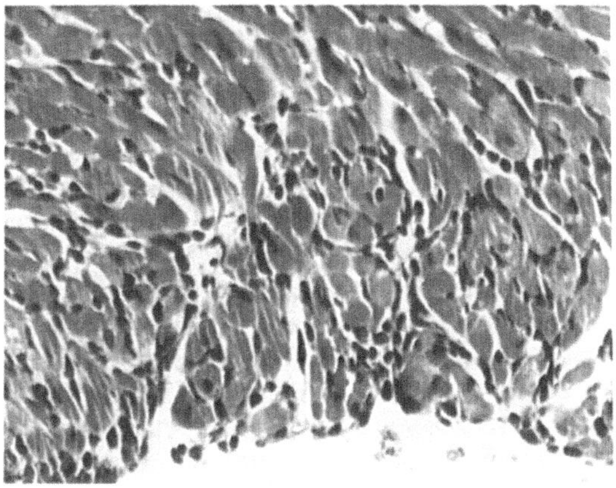

Fig. 2.6. Medium power photomicrograph taken from the interventricular septum of an A/J mouse 15 days after infection with CB3. Scattered interstitial mononuclear inflammatory cells spread in a spider-like fashion, between myocytes that appeared normal. 440 ×.

Even 45 days after infection, there was continued inflammation with persistent interstitial infiltration by mononuclear cells. During the late phase of myocardial disease, heart-specific autoantibodies were detected in the sera of susceptible animals. The autoantibodies showed both the sarcolemmal and interfibrillar patterns of reaction in indirect immunofluorescence tests, resembling the appearance obtained using human sera.

The two cardinal features of the late phase of disease are the presence of the diffuse mononuclear infiltration and heart-reactive autoantibodies [15]. The transition from early to late phase myocarditis begins on approximately day 9. The pathological changes were evident only in the heart; no inflammatory lesions were found in skeletal muscle, pancreas or other tissues. To test the specificity of the heart-reactive autoantibodies, absorption experiments were carried out [16]. Lyophilized skeletal muscle removed some, but not all, of the heart-reactive autoantibody whereas heart tissue removed all of the antibody. We concluded, therefore, that two populations of antibodies were present in the late phase of myocarditis. One type of antibody reacted both with cardiac and skeletal muscle whereas a second antibody was specific for heart muscle.

Experiments were undertaken to test the role of MHC and non-MHC encoded genes in the late phase of virus-induced myocarditis. Four H-2 congenic panels of mouse strains were compared with respect to their pathology indices. All four congenic strains on the A background developed typical late phase disease but the lesions were significantly more severe in A/J (H-2^a) and A.BY (H-2^b) than they were in A.CA (H-2^f) and A.SW (H-2^s) mice. Since these animals are all genetically identical except for the small segment of chromosome 17 on which the MHC is located, these effects are attributable to H-2. A non-H-2 effect could also be discerned from these studies. Five strains shared the H-2^b haplotype; three strains, A.BY, BALB/b, and C3H.SW, developed late phase myocarditis as evidenced by the typical interstitial infiltrate and the presence of heart-specific autoantibodies. Two strains, C57BL/6 and C57/BL10, did not develop late disease. Therefore, non-MHC genes contribute to late phase disease [17].

In order to localize the non-MHC sites of control, a series of recombinant inbred (RI) lines developed by Dr M. Nesbit was studied [18]. The animals were infected with CB3 in the usual manner, and the prevalence of myocarditis and circulating heart-specific antibodies was determined on day 21. By using RI lines derived from A/J (susceptible) and C57BL/6 (resistant) parental lines, one or more genes influencing the autoimmune disease were located on chromosome 14. The trait(s) showed linkage to the genes controlling the alpha chain of the T cell receptor. Since the T cell receptor determines the first step of immunological recognition, it might well encode for response to the relevant immunopathic antigen. Further experiments were designed to determine the mode of inheritance of the heart autoimmunity. When crosses were carried out between two A strain congenics that differed only at the H-2 region, autoimmune myocarditis was dominant. On the other hand, when non-H-2 differences were studied the F1 hybrids were poor responders, suggesting that an autosomal recessive gene is involved in susceptibility to autoimmune myocarditis [17].

The availability of post-infection autoimmune sera from mouse strains that develop the late phase of myocarditis permitted us to define the antigen responsible for autoimmune response [19]. Antiserum pools were prepared from susceptible and non-susceptible strains of mice and tested with various extracts prepared from normal mouse hearts. The extracts were separated by sodium dodecyl sulphate polyacrylamide gel electrophoresis and the constituent proteins transblotted to nitrocellulose. The proteins were then detected by treatment with the respective antiserum pools. Serum samples collected from both susceptibles and non-susceptibles on days 5 and 7 after infection exhibited staining of a weakly-staining band at 200 kDa, the molecular weight of myosin heavy chain. A similar pattern was observed with normal serum, and when skeletal muscle extracts were tested. On the basis of these results we concluded that normal serum and serum from animals with early phase disease contain antibodies to myosin. Using serum pools collected 15–21 days after infection from resistant strain mice, we found a similar staining pattern with both heart extract and skeletal muscle extract. However, when the serum pool from the mice susceptible to the late phase of myocarditis was tested a pronounced reaction was seen with the 200 kDa band. We found that this reaction could be removed by absorption with cardiac myosin, but not with skeletal muscle myosin. Moreover, antibody eluted from immobilized cardiac myosin strongly stained the 200 kDa antigen. From these experiments we concluded that an important antigenic target of post-infectious autoimmune myocarditis in susceptible strains of mice is the heavy chain of cardiac myosin.

These results raised the possibility that autoimmunity to cardiac myosin might be responsible for post-infectious myocarditis. We tested this possibility by immunizing genetically susceptible and resistant strains of mice with purified murine cardiac myosin [20]. In addition, we used skeletal muscle and brain myosin for comparison. Serum collected 15–21 days after immunization showed IgG specific for cardiac myosin only in animals that were immunized with cardiac myosin. Furthermore, these animals developed typical lesions of late phase myocarditis. From these experiments we concluded that immunization of genetically susceptible strains with cardiac myosin induces a response analogous to that produced by CB3 infection.

Myosin-Induced Myocarditis in Mice

In order to carry out a detailed comparison of myosin-induced myocarditis and CB3-induced myocarditis parallel experiments were conducted in groups of genetically susceptible A/J mice. Antibody to myosin was evident in the myosin-immunized mice on day 15, a time when myosin antibody is not yet present in the CB3 infected animals (Fig. 2.7) By day 28, however, both CB3 and myosin-immunized animals had IgG antibody to cardiac myosin as measured by western immunoblots (Fig. 2.8). Animals of each group were sacrificed at that time and found to have moderate to severe myocarditis; the histological features of myosin-induced autoimmune myocarditis are given in Fig. 2.9. The picture should be compared with the late phase of CB3-induced disease shown in Fig. 2.6. When immunocytochemical techniques were applied to the hearts of myosin-immunized mice, deposits of IgG and C3 could clearly be discerned on the myocyte cell surfaces (Figs. 2.10 and 2.11).

These results clearly show that autoimmune myocarditis can be produced in genetically susceptible mice by two means: CB3 infection and myosin immunization. The histological features of the two forms of the disease are strikingly similar. There are, however, some significant differences in that the virus-induced form of autoimmune myocarditis may be associated with the presence of additional autoantibodies as suggested by the tissue staining patterns by immunofluorescence. We can say with certainty, then, that myosin is a principal target of the autoimmune response in CB3-infected mice.

These investigations raise a number of questions about the association of coxsackie virus infection and myocarditis. We first considered the possibility that the virus shares antigenic determinants with myosin. Using high titred antisera from CB3-infected animals, however, we were unable to demonstrate any reaction with the viral capsid proteins. Moreover, the tempo of neutralizing

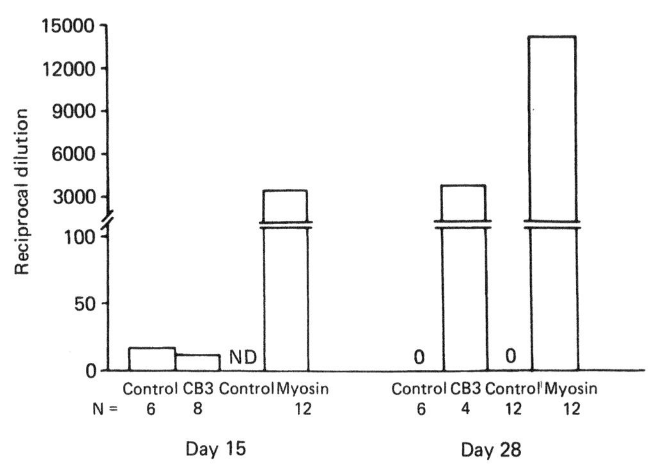

Fig. 2.7. Mean titre of circulating antibodies (IgG) to murine cardiac myosin in A/J mice infected with CB3 or immunized with murine cardiac myosin 15 and 28 days after treatment. Mean titres are given for animals receiving the appropriate control treatments. Titres were determined by enzyme-linked immunosorbent assay (ELISA). The number of animals from each group tested ranged from 4 to 12; no data (*ND*) were available for myosin treatment controls on day 15.

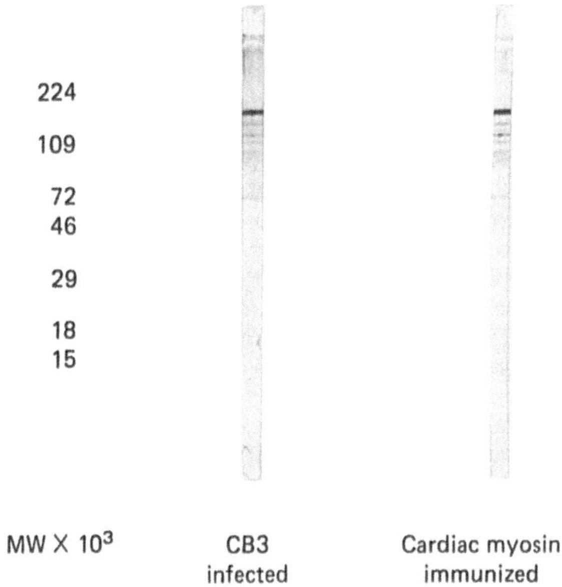

Fig. 2.8. Western immunoblots showing cardiac myosin reactive antibodies in serum from A/J mice 28 days after infection with CB3 or immunization with murine cardiac myosin. The sera were incubated with immobilized extracts of normal mouse hearts and reactive antibodies detected by an amplified horseradish peroxidase signal specific for mouse IgG.

antibody production to the virus differs greatly from the kinetics of myosin-antibody production. Neutralizing antibody appears at day 2 or day 3 with the highest titre appearing by day 9. Antibody to myosin does not appear until day 9. For these reasons we have proposed the alternative possibility that endogenous myosin is the antigenic stimulus. Here the problem arises that myosin is considered to be an internal constituent of the cell and not available to initiate an

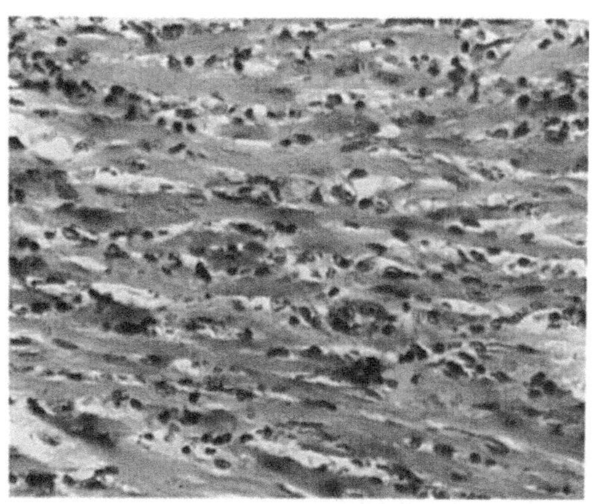

Fig. 2.9. Medium power photomicrograph taken from the left free wall of an A/J mouse 15 days after immunization with murine cardiac myosin. See Fig. 2.6 to compare with the interstitial inflammation observed 28 days after CB3 infection. 440 ×.

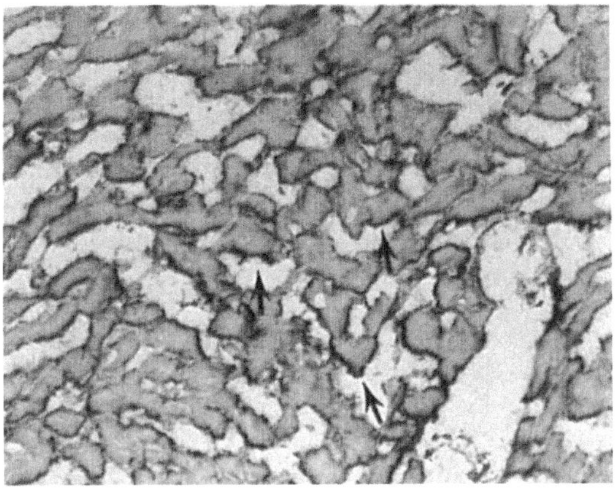

Fig. 2.10. Medium power photomicrograph of in situ deposition of antibody (IgG) in the left ventricular free wall of an A/J mouse 28 days after immunization with murine cardiac myosin. Note the diffuse staining of the sarcolemmal membranes of all the myocytes (*arrows*). Antibody deposition was detected by direct staining with a horseradish peroxidase-labelled second antibody specific for mouse IgG. 440 ×.

immune response. Figure 2.9 suggests, however, that determinants of myosin are available on the myocyte surface. Alternatively, virus-induced tissue damage may release myosin into the intercellular spaces. Since myosin is a relatively large, insoluble molecule, it is likely to remain exposed for a sufficient period of time to initiate an immune response. In addition, the virus attracts inflammatory cells that are capable of processing and presenting myosin and providing the interleukin signals necessary to initiate antibody production. Whether post-

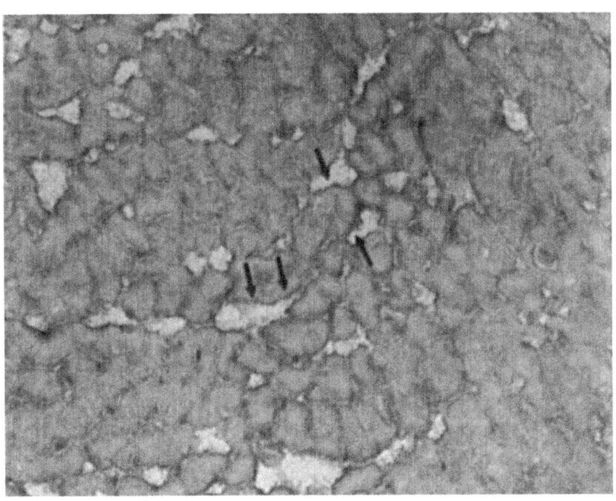

Fig. 2.11. Medium power photomicrograph of in situ deposition of complement (C3) in the left ventricular free wall of an A/J mouse 28 days after immunization with murine cardiac myosin. This tissue section was obtained from the same animal that was used for Fig. 2.10. Note a similar diffuse staining of the sarcolemmal membranes (*arrows*). Complement deposition was detected by direct staining with a horseradish peroxidase-labelled second antibody specific for the C3 complement component. 440 ×.

infection myocarditis is an example of autoimmunity due to molecular mimicry or to exposure of intracellular constituents is not clearly resolved.

There are marked differences in the course of disease due to the strain of virus used. All the experiments described above were carried out with the Nancy strain of CB3 obtained from Dr Martin Lerner, Wayne State University. Gauntt et al. [21] have carried out extensive investigations with different strains of CB3 and found that they produce strikingly different responses in various strains of inbred mice.

Implications for Human Myocarditis

Studies of autoimmune myocarditis in the mouse have provided fresh insights into human cardiomyopathy and myocarditis. An important first step in these studies was to distinguish the early stage of the disease during which the infectious agent is present from the later stage when no infectious micro-organisms can be isolated. In the case of CB3-induced murine myocarditis this first step, lasting about 9 days after infection, depends upon a number of genetic factors of the host, especially those influencing the tempo of neutralizing antibody production. The strain of virus is probably of equal importance. The pathological changes are most likely attributable directly to the presence of virus and the inflammatory response that it evokes. If the mouse model is an accurate indicator, this early stage of disease in the human probably heals with little or no residuum in most patients. In fact, the patient is likely to be unaware that the myocarditis process is occurring unless the heart is severely stressed.

The late stage of myocarditis represents an autoimmune response to the affected organ. Several properties of the autoimmune response may make it difficult to recognize in humans. First, it appears to be relatively rare, occurring in genetically predisposed individuals. Second, it may be remote in time from the original infection; in the human the immunopathic stage of disease may occur months after the initial infection. Third, the infectious agent is difficult to demonstrate in the heart. New methods of nucleic acid hybridization may make it possible to find evidence of the micro-organism, even if it is not present in an infectious form. Fourth, the autoimmune response may be complex. In the mouse the primary antigen involved in post-coxsackie myocarditis is the cardiac isoform of myosin. It is likely, however, that a number of additional autoantibodies to heart are present in these infected animals. In the human the autoantibody pattern is even more complex. It may well be that myosin is not the principal autoantigen; at the moment, however, myosin is the only cardiac antigen known to produce myocarditis in experimental animals. Finally, the diagnosis of post-infectious myocarditis is complicated by the fact that many humans have naturally occurring autoantibody to myosin.

A better understanding of post-infection autoimmune disease has a number of important clinical implications. First, it will enable us to identify biomarkers of susceptibility. In the mouse the H-2 haplotype correlates with severity of the autoimmune response, but susceptibility resides primarily in non-H-2 background genes. The presence of autoantibodies may be an important adjunct to the diagnosis of post-infectious autoimmune disease. Such tests, however, must be applied with caution since autoantibodies to normal cell components are commonly found; attention must be given both to the titre and isotype of antibody. The value of autoantibodies for diagnosis is based on empirical association; there is no implication that the antibodies are necessarily responsible for the pathology of the disease. In the case of myocarditis, for example, it is possible that the actual tissue injury is attributable to cell-mediated immunity. The presence of autoantibodies, then, is an indication that an autoimmune response is underway. Finally, a better understanding of post-infectious autoimmunity will be a surer guide to future therapy. In the case of myocarditis there has long been debate about the value of immunosuppressive treatment. Logically, immunosuppression is useful in stages of the disease when active virus infection is no longer occurring and the ongoing lesions are due to autoimmunity. The mixed reports of the value of immunosuppression in myocarditis and dilated cardiomyopathy may be due to the fact that such treatment is expected to be beneficial in patients whose disease is immunologically mediated [22].

There are many lessons still to be learned. It is important, we feel, to dissect the post-infection autoimmunities into their stages by appropriate use of experimental models. By manipulating the genetics of the host as well as the virus, it is possible to distinguish the different phases of disease; only then can the pathological mechanisms of each be worked out.

Summary

The role of viruses in inducing autoimmune disease has been debated for many years. We have developed a model of autoimmune myocarditis in weanling mice produced by infection with a myocarditogenic strain of coxsackie virus B3. The consequent disease can be resolved into two phases. Early myocarditis, seen on days 3–9 after infection, is characterized by focal necrosis and calcification; infectious virus can still be recovered. In most mouse strains, this disease resolves spontaneously. A late phase of myocarditis develops in a few, genetically susceptible, strains. It is characterized by diffuse mononuclear infiltration and production of heart-specific autoantibodies. No infectious virus can be recovered during the late phase of disease. The autoantibodies are specific for the cardiac isoform of myosin heavy chain. Purified murine myosin induces heart-specific autoantibodies and typical diffuse myocarditis in genetically susceptible mice. Genetic susceptibility to autoimmune myocarditis is a polygenic trait involving both MHC and background genes. This example of virus-induced autoimmunity may be useful in analysing the role of micro-organisms in autoimmune disease.

References

1. Helin M, Savola J, Lapinleimu K (1968) Cardiac manifestations during a coxsackie B5 epidemic. Br Med J 3:97–99
2. Cambridge G, MacArthur CGC, Waterson AP, Goodwin AF, Oakley CM (1979) Antibodies to coxsackie B viruses in congestive cardiomyopathy. Br Heart J 41:692–696

3. Kandolf R, Hofschneider PH (1989) Enterovirus-induced cardiomyopathy. In: Notkins AL, Oldstone MBA (eds) Concepts in viral pathogenesis III. Springer, Berlin Heidelberg New York, pp 282–290
4. Maisch B, Deeg P, Liebau G, Kocksiek K (1976) Diagnostic relevance of humoral and cytotoxic immune reactions in primary and secondary dilated cardiomyopathy. Am J Cardiol 52:1072–1078
5. Maisch B, Berg PA, Kochsiek K (1980) Autoantibodies and serum inhibition factors (SIF) in patients with myocarditis. Klin Wochenschr 58:219–225
6. Maisch B, Trostel-Soeder R, Stechemesser E, Berg PD, Kochsiek K (1982) Diagnostic prevalence of humoral and cell-mediated immune reactions in patients with acute viral myocarditis. Circulation 70:149–156
7. Scheerder I de, Vandekerckhove J, Robbrecht J et al. (1985) Post-cardiac injury syndrome and an increased humoral immune response against the major contractile proteins (actin and myosin). Am J Cardiol 56:631–633
8. Das SK, Callen JP, Dodson VN, Cassidy JT (1971) Immunoglobulin binding in cardiomyopathic hearts. Circulation 44:612–616
9. Schultheiss HP, Bolte HD (1985) Immunological analysis of autoantibodies against the adenine nucleotide translocator in dilated cardiomyopathy. J Molec Cell Cardiol 17:603–617
10. Woodruff JF (1980) Viral myocarditis: a review. Am J Pathol 101:425–484
11. Reyes MP, Ho KL, Smith F, Lerner AM (1981) A mouse model of dilated-type cardiomyopathy due to coxsackievirus B3. J Infect Dis 144:232–236
12. Wolfgram LJ, Beisel KW, Herskowitz A, Rose NR (1986) Variations in the susceptibility to coxsackievirus B_3-induced myocarditis among different strains of mice. J Immunol 136:1846–1852
13. Herskowitz A, Wolfgram LJ, Rose NR, Beisel KW (1987) Coxsackievirus B_3 murine myocarditis: A pathologic spectrum of myocarditis in genetically defined inbred strains. J Am Coll Cardiol 9:1311–1319
14. Beisel KW, Wolfgram LJ, Herskowitz A, Rose NR (1985) Differences in severity of coxsackie virus B_3 induced myocarditis among H-2 congenic mouse strains. In: Skamene E (ed) Genetic control of resistance to infection and malignancy. Alan R. Liss, New York, pp 195–201
15. Rose NR, Wolfgram LJ, Herskowitz A, Beisel KW (1986) Postinfectious autoimmunity: two distinct phases of coxsackievirus B_3-induced myocarditis. Ann NY Acad Sci 475:146–156
16. Wolfgram LJ, Beisel KW, Rose NR (1985) Heart-specific autoantibodies following murine coxsackievirus B_3 myocarditis. J Exp Med 161:1112–1121
17. Rose NR, Neumann DA, Herskowitz A (1988) Genetics of susceptibility to viral myocarditis in mice. Pathol Immunopathol Res 7:266–278
18. Traystman M (1988) Genetic regulation of CVB3-induced autoimmune myocarditis. Dissertation, Johns Hopkins University, Baltimore MD
19. Alvarez FL, Neu N, Rose NR, Craig SW, Beisel KW (1987) Heart-specific autoantibodies induced by coxsackievirus B_3: Identification of heart autoantigens. Clin Immunol Immunopathol 43:129–139
20. Neu N, Rose NR, Beisel KW, Herskowitz A, Gurri-Glass G, Craig SW (1987) Cardiac myosin induces myocarditis in genetically predisposed mice. J Immunol 139:3630–3636
21. Gauntt CJ, Gomez PT, Duffey PS et al. (1984) Characterization and myocarditic capabilities of coxsackievirus B3 variants in selected mouse strains. J Virol 52:598–605
22. Parrillo JE, Cunnion RE, Epstein SE et al. (1989) A prospective, randomized, controlled trial of prednisone for dilated cardiomyopathy. N Engl J Med 321:1061–1068

Chapter 3

The Role of Environmental Factors in Modulating Insulin Dependent Diabetes

E. H. Leiter

Introduction

Considerable evidence indicates that insulin dependent diabetes mellitus (IDDM) has an autoimmune aetiology [1]. This evidence includes the frequent detection of islet cell surface and cytoplasmic autoantibodies in the prodromal period, as well as IDDM occurrence in patients with polyendocrine autoimmune disorders such as systemic lupus erythematosus, Hashimoto's thyroiditis, and Graves' disease. Perhaps the most compelling evidence for the autoimmune nature of IDDM in humans has been the fate of segmental pancreatic grafts transplanted from a normal sib into a monozygotic twin with IDDM of long duration. In the absence of immunosuppression, only a transient amelioration of hyperglycaemia was observed with islets in the graft developing insulitis and selective destruction of the islet β cells within 2 months [2]. The strongest genetic susceptibility modifiers are tightly linked to the HLA-DR and DQ loci, and could, in fact, represent one or more of these genes themselves [3,4]. Concordance rates as high as 54% in monozygotic twins have confirmed the importance of genetic factors in aetiopathogenesis of IDDM, but the lack of 100% concordance suggests that environmental factors (pathogenic micro-organisms and viruses, toxins or additives in the diet) may modify penetrance of diabetes susceptibility genes [5]. The contribution of environmental factors in modulating the penetrance of susceptibility genes has remained unresolved, and indeed has been the subject of some controversy [6]. Monozygotic twins represent the classic system for evaluating the interplay between heredity and environment in the aetiopathogenesis of human constitutive disorders. In IDDM, less than 50% concordance of IDDM in monozygotic twin studies suggests that genetically-inherited susceptibility alone is insufficient to elicit disease. Clear evidence exists in the literature to document the presence of environmental "triggers", either in the form of diabetogenic viruses [7], environmental toxins of microbial origin [8], or factors in the diet [9]. However, the absence of perfect concordance in monozygotic twin studies need not reflect differential exposure to diabetogenic initiators in the environment [10]. Although identical twins inherit the same germ line genes (and are MHC-identical), random recombinatorial rearrangements occur to produce functional T cell receptor genes and B lymphocyte immunoglobulin genes necessary for the constitution of an immune system [11]. If only certain recombinatorial T cell receptor gene rearrangements can confer ability to recognize and react against β cell specific autoantigens, then lack of 100% concordance could be explained by the stochastic process of diversity generation [10]. Oligoclonal use of $V_{\beta}8$ genes by encephalolitigenic T cells has been demonstrated in certain inbred strains of mice susceptible to experimental allergic encephalomyelitis induced by injection of myelin basic protein [12]. Even if an immune system capable of mediating autoimmune reactivity against β cells in an individual was the product of chance recombinations of hypervariable genetic loci, additional accelerative stress mediated through environmental agents may still be required to achieve full penetrance of the genomic susceptibility modifiers. Environmental factors such as dietary toxins of microbial origins and pathogenic viruses clearly pose the potential for destructively interacting with a diabetes-susceptible genotype not only by compromising β cell viability and function, but also by significantly altering β cell surfaces in such a way as to elicit an immune-mediated cytotoxic response. Epidemiological evidence suggests that there is a strong dietary component in the initiation of IDDM; for example, when Polynesian populations exhibiting low IDDM incidence migrate to more advanced countries (where milk and wheat proteins are plentiful in the diet as compared with the island of origin), the incidence of IDDM approaches that of the host country [6]. However, in humans, rigorous demonstration that strong interactions between environmental stresses and an IDDM-susceptible genotype "force" the penetrance of IDDM susceptibility modifiers has been difficult, partly because of the near impossibility of identifying and recruiting suitably matched groups of prediabetic individuals who can be forced to consume defined diets or be exposed to an environmental pathogen over a period of years and under conditions in which other environmental variables are held constant. In this regard, laboratory rodents spontaneously developing autoimmune IDDM syndromes present ideal subjects.

Insights from Animal Models

The BioBreeding Rat

BioBreeding (BB) rats spontaneously develop autoimmune IDDM, and thus provide an excellent test of the interaction between genetics and environment. Even in this model of heritable IDDM, diabetes incidence in any generation seldom reaches 100% and usually varies between 40% and 80%, depending on the colony [13]. Most, but not all, BB rats exhibit a permanent lymphopenia and lymphocytopenia and thus are immunodeficient throughout life. Non-lymphopenic substrains have been produced, and one of them is diabetes-resistant [14]. In rare cases where diabetes spontaneously occurs in non-lymphopenic substrains, a different pathogenetic mechanism is suggested by the development of an earlier onset diabetes, and by evidence suggesting the involvement of different immune effector cell populations than those present in the lymphopenic strains [15]. Diabetogenesis is under polygenic control, with requirement for at least one RT-1^u haplotype (with the tightest association to the class II loci) as well as a completely recessive, non-MHC linked gene controlling lymphopenia [14,16]. This lymphopenia gene is strongly penetrant, and thus is not likely to be the same 50% penetrant recessive diabetes gene proposed in another study [17]. That multiple control loci are involved is illustrated by outcross with unrelated rat strains; diabetes incidence in F2 generations is usually only 3%–5% [16], whereas lymphopenia is present at approximately 25% incidence [14].

The question of environmental factors influencing penetrance of diabetogenic susceptibility genes in lymphopenic BB/Wor rats was initially tested by Caesarean-derivation of pups into a germ-free (gnotobiotic) environment [18]. Of 12 gnotobiotically-reared rats, 2 males and 1 female became diabetic (25% incidence). The conclusion was that penetrance of the susceptibility genes was not dependent upon absence of specific pathogens. Further evidence indicated that the diabetogenic susceptibility was not responsive to manipulations in either dietary macronutrients or to sex steroids [19], distinguishing this model from autoimmune lupus-prone strains of mice such as NZB and NZW [20,21]. The diet study entailed increases in the carbohydrate, protein, or lipid macronutrient content of a chow-based diet [19]. Analysis of the diabetogenic agents in permissive chow-based diets (complex mixtures of natural ingredients) by addition of components into diabetes-retardant semi-purified diets such as AIN-76 indicate the likelihood of complex interactions between dietary proteins and unknown passenger xenobiotics or microbial components [22]. The potential nature and origins of such diabetogenic catalysts will be discussed below. Even when feeding was initiated at weaning, therapeutic AIN-76 diets failed to block development of insulitis in BB rats, and none of the therapeutic diets has been demonstrated to improve the immunodeficiencies [23]. Unlike certain other autoimmune prone models [24], feeding BB rats a semi-purified diet supplemented with fish oil conferred no added protection against diabetes in comparison to a safflower-oil-supplemented diet [22]. Safflower oil and corn oil (the latter represents the lipid component in semi-purified diet AIN-76) are enriched in *n*-6 fatty acids, primarily linoleic acid. In contrast to fish oils rich in *n*-3 fatty acids, these linoleic-enriched vegetable oils tend to increase cellular arachidonic acid content. Increased arachidonic acid metabolism could lead in turn to perturbations in immunological, as well as pancreatic islet secretory functions through elevated synthesis of eicosanoids, including prostaglandins, thromboxanes and leukotrienes [25,26]. Macrophages (Mϕ) are central to pathogenesis in this model [27]. Not surprisingly, therefore, feeding an essential fatty acid deficient diet that prevented Mϕ migration into the islets and depleted basal concentrations of resident leukocytes in the islets, also markedly protected BB/W rats [28].

Recent studies entailing experimental infection of BB rats with pathogenic murine viruses have demonstrated that diabetogenesis in this model is also modifiable by microbial challenges from the environment. Inoculation of lymphocytic choriomeningitis virus (LCM, clone 13) into prediabetic BB rats significantly reduced the prevalence of diabetes, presumably by altering function of the reduced numbers of peripheral T cells – either by stimulating regulatory T cells capable of suppressing the generation of cytotoxic effectors or by deleting the effectors directly [29]. Similarly, treatment of BB rats with OK 432, a treated streptococcal suspension, also circumvented diabetes [30]. Thus, in the BB model, it has been demonstrated that environmental factors are capable of affecting the penetrance of genetic susceptibility modifiers.

The Non-obese Diabetic Mouse

Non-obese diabetic (NOD) mice, like BB rats, spontaneously develop autoimmune IDDM [31], but unlike BB rats, NOD mice are neither T-lymphocytopenic nor T-lymphopenic [32]. Diabetogenesis in NOD/Lt mice is polygenetically controlled and is mediated by pancreatic β cell-specific immune effectors derived from bone marrow [33]. The diabetogenic susceptibility modifiers include a minimum of three recessive (one MHC-linked) plus some dominant susceptibility genes [32]. These modifiers predispose to development of a dysregulated immune system. Mφ from NOD/Lt mice present soluble antigens to T helper cells very well [33], yet the unique MHC class II products of NOD fail to activate T suppressor (Ts) cells in a syngeneic mixed lymphocyte reaction (SMLR) [34]. The defective SMLR in NOD/Lt mice has been linked to the MHC class II bearing stimulator cell population, and involves deficiencies in cytokine signalling between Mφ and T lymphocytes [34,35]. These deficiencies include subnormal interleukin-1 (IL-1) and interleukin-2 (IL-2) production by leukocytes in vitro; deficiencies in high affinity IL-2 receptors on NOD T cells have also been reported [36]. In addition, NOD/Lt mice develop an abnormally elevated ratio of "cytotoxic/suppressor" $CD8^+$ T lymphocytes as a percentage of total T lymphocytes [32].

Environmental Modulation of Penetrance of Diabetogenic Susceptibility Genes in NOD Mice

Environmental Factors as Biological Response Modifiers

NOD mice present a striking illustration of the powerful influence that environmental pathogens exert in modulating penetrance of diabetogenic genes. The relationship between environmental immunostimulation and diabetogenesis is best illustrated by a consideration of the extreme variation in diabetes prevalence in males from various colonies of NOD mice. The NOD strain was developed by Makino at the Shionogi (Shi) Laboratories, Japan [31]. NOD/Shi mice in this source colony exhibited a strong gender bias for diabetes susceptibility, with 80% of the females, but less than 20% of the males, becoming diabetic by 210 days [31]. This gender bias in females was reminiscent of that seen in autoimmune lupus-prone strains of mice, and was consistent with the observation that female mice generally mount stronger immunological responses than do males. Since no such female gender bias characterizes either the BB rat model, or type 1 diabetes in humans, this was seemingly a characteristic unique to the genus *Mus*. That this did not represent an intrinsic species difference, but rather an unusual sensitivity to environmental influences, was demonstrated when colonies of NOD mice were established outside of Japan. The NOD/Lt colony maintained by the author at The Jackson Laboratory (TJL) originated from a shipment of NOD/Shi breeders to the Joslin Diabetes Center (Boston, MA). The shipment was divided such that two breeding pairs of NOD were sent directly to TJL and the remainder went into the animal facilities at Joslin (Jos) which, at the time, were not free of all murine pathogenic viruses. Upon receipt at TJL, the mice were bred in a quarantine facility and their offspring were Caesarean-derived and fostered to specific pathogen-free (SPF) DBA/2J dams in isolator cages. Upon verification that the Caesarean-derived mice were free of murine pathogenic viruses and mycoplasmas, they were transferred into a SPF research mouseroom for further inbreeding and establishment of diabetes incidence.

A critical question was how penetrant the diabetogenic susceptibility genes would be in the TJL colony. Diabetes incidence in NOD/Lt females was as high as in Japan, with between 80% and 95% diabetes incidence by 40 weeks. It was initially assumed that males would not be useful for analysis in genetic linkage studies because of the low penetrance of diabetogenic genes reported for NOD/Shi males [31]. However, a striking difference in the incidence of diabetes in NOD/Lt males and NOD/Shi mice, as well as the NOD/Jos maintained by Dr M. Hattori at the Joslin Diabetes Center, was observed. Whereas the incidence in these latter two colonies was less than 20% by 30 weeks, NOD/Lt males were exhibiting incidences of around 50%. By 52 weeks, the NOD/Lt male incidence (>70%) approached the 90% incidence observed in NOD/Lt females. Further evidence that the environment was influencing penetrance of diabetogenic genes was furnished when NOD/Lt breeders were shipped to Dr B. Singh in Edmonton, Canada. A 100% diabetes prevalence in NOD/Lt mice of both sexes by 186 days was obtained in Dr Singh's modern facilities at the University of Alberta/Edmonton (personal communication). Indeed, since Dr Makino's original report of a low male diabetes incidence in his conventional SPF colony, he has rederived his breeding stock by egg transfer into pseudopregnant carriers within a newly fumigated mouseroom at the Shionogi Aburahi Laboratories, Japan. Since these manipulations, he has observed a rise in male diabetes prevalence to a current incidence of around 40% (S. Makino, personal communication). In another study in Japan, diabetes incidence in NOD males raised in a conventional colony was compared with that in males raised gnotobiotically in isolator cages. The results confirmed that environmental microbial agents may actually protect NOD mice from diabetes incidence since the 6% incidence in the conventional SPF colony increased to 71% under germ-free conditions [37]. The SPF status of the colony is not the only environmental influence determining the penetrance of the diabetes susceptibility genes. At the same time that NOD/Lt breeding pairs from TJL were sent to Dr Singh in Edmonton, breeding pairs were also sent to Dr Aleyne Ricker, East Carolina State University, North Carolina. Upon receipt, the mice were placed in micro-isolator cages and maintained on a semi-purified diet. No diabetes was observed in this colony until the diet was changed; following replacement of this semi-purified diet with a natural ingredient Purina diet, diabetes incidence returned to the level observed in NOD/Lt at TJL (Dr A.

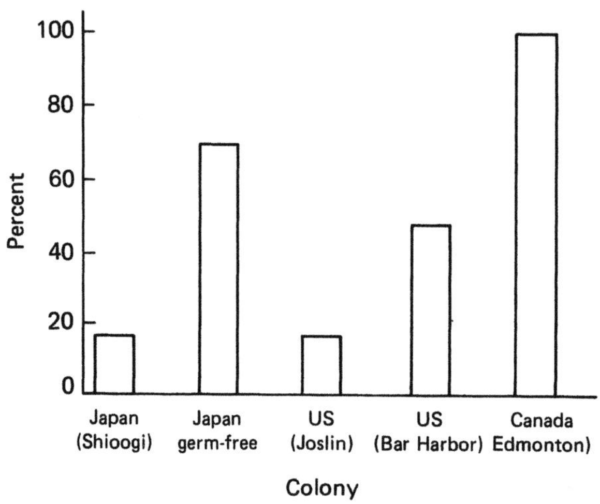

Fig. 3.1. Colony variation in the diabetes incidence (percent cumulative diabetes) in NOD male mice.

Ricker, personal communication). An illustration of environmental effects on diabetes penetrance is shown in Fig. 3.1.

Viruses

In humans, circumstantial evidence indicates that exposure to pathogenic viruses may represent one of the environmental "triggers" that, in conjunction with a susceptible genotype, initiate a chain of events culminating in the generation of an autoimmune attack against the β cells [38]. However, the results described above indicate that the more stringent the SPF husbandry conditions, the greater the penetrance of NOD diabetogenic susceptibility genes. This would indicate that in the NOD model, immunostimulation produced by response to a viral infection would be protective. The first confirmation of this was obtained by Oldstone, who circumvented diabetes in NOD mice by experimental infection with the Armstrong clone 13 strain of lymphocytic choriomeningitis virus (LCM) [39]. It will be recalled that infection by this virus also protected the immunodeficient BB rat [29]. Although LCM virus is lymphotropic, the protection in NOD was apparently not due to generation of an immunodeficient state since lymphocytes could be recovered from LCM-infected mice that failed to release infectious virus. Thus, transfer of these cells into irradiated NOD mice was not associated with LCM infection of the recipients. In fact, lymphocytes from LCM-infected (diabetes-protected) donors adoptively transferred protection from diabetes as opposed to overt diabetes [39]. Apparently, the immune system is in some way "distracted" from developing anti-β cell immunity when challenged by extrinsic immunologic stimuli. Failure to activate functional T-suppressor networks in uninfected NOD mice may reflect defective cytokine release [34]; if so, strong non-specific immunostimulation may protect from diabetes by eliciting a more robust cytokine release from both Mϕ and lymphocytes, in turn stimulating a more normal immunoregulatory (T-suppressor?) network. The issue of cytokine production in response to stimulation by microbial challenge will be discussed in more detail below.

Although diabetes protection afforded by LCM virus infection is beneficial to NOD mice, loss of diabetes prevalence in an NOD colony resulting from an uncontrolled exposure to an environmental pathogen seriously confounds interpretation of experimental data. Over a span of four inbreeding generations, diabetes incidence in an NOD colony at the Diabetes Research Institute, Düsseldorf, West Germany, plummeted from 80% incidence in females to 1%; serological screening indicated that infection by murine hepatitis virus was responsible for the loss of diabetes penetrance (Dr L. Herberg, personal communication). In such a situation, the investigator's only recourse is to destroy the infected colony, decontaminate the facilities, and begin anew with clean breeding stock. Protection of a colony not only requires routine monitoring for absence of murine pathogenic agents in the mice themselves, but also requires the screening of all reagents of rodent origin, such as monoclonal antibodies, used for studies in vivo. It is interesting that in BALB/cByJ males treated with multiple low doses of streptozotocin (MSZ) to induce IDDM, viral exposure had the opposite effect than would be expected on the basis of the NOD studies. Prior to the complete eradication of an enzootic viral contaminant (pneumonia virus of mice, PVM), BALB/cByJ males developed hyperglycaemia within 2 weeks of the MSZ treatment. This hyperglycaemic response was strongly attenuated after the colony was rederived to eliminate any PVM carriers [40]. MSZ-damaged islets become infiltrated with Mϕ which secrete IL-1, tumour necrosis factor (TNFα), and interferons (IFNs). These cytokines exert strong synergistic cytotoxic effects [41] and are therefore capable of compounding MSZ-induced β-cytotoxicity. This is illustrated by the development of diabetes in ICR Swiss or CD-1 mice first treated with a subdiabetogenic dose of streptozotocin (SZ), followed by challenge with the non-diabetogenic B variant of the encephalomyelocarditis (EMC) virus, a strong interferon inducer [42]. The host response to the viral stimulus must follow the single low SZ dose since no hyperglycaemia is produced if the virus challenge precedes the sub-diabetogenic SZ dose. Thus, in this model, the cytokines elicited by an infection with a non-diabetogenic virus very likely compromise the ability of surviving (but SZ-damaged) β cells to maintain glucose homeostasis. Viral infections in NOD mice have a different consequence. In the NOD model, a clear autoimmune reactivity against β cells occurs spontaneously as a consequence of a dysregulated immunoregulatory network. This dysregulation appears to be the result of deficiencies in cytokine signalling. Hence, systemic increases in monokine and lymphokine secretions elicited by viral challenge actually protect NOD mice from hyperglycaemic development, most likely by elevating endogenous cytokine levels and thereby boosting immunoregulatory pathways capable of retarding the activation of cytotoxic T-effector cells. At TJL, animal health strictures preclude the experimental infection of NOD/Lt mice with pathogenic viruses. We have instead simulated viral infection by treating prediabetic females with polyinosinic-polycytidylic acid [poly I:C], a potent inducer of IFM α/β. This treatment completely circumvented diabetes, apparently through activation of a population of natural suppressor cells [43]. Immunological responses to persistent microbial infections may result in the activation of natural suppressor cells by stimulating macrophages to produce IFN α/β in a manner similar to poly I:C treatment.

Bacterial Products

Bacteria, and especially their cell wall antigens, are potent biological response modifiers. Injection of bacterial suspensions into prediabetic NOD mice prevents diabetes onset. A single injection of complete Freund's adjuvant (emulsified mycobacteria), or weekly injections of OK 432 (a lyophilized streptococcal suspension) provided long-term diabetes protection [44,45]. The latter compound is a potent activator of Mϕ and stimulates monokine secretion [46]. We have demonstrated that the NOD stimulator cell defect in an SMLR can be overcome by supplementation of either IL-1 or IL-2 [34]; more recently, we have demonstrated that chronic treatment of prediabetic NOD females in vivo with low doses of IL-2 not only restored normal SMLR activity, but circumvented diabetes [43]. Unpublished studies from other laboratories have shown that chronic treatment with IL-1 and TNFα also circumvents diabetes. These pharmacological studies demonstrate that all of the components for a normal immunoregulatory network are present in NOD mice, and that environmental immunostimulation is required to maintain function. In this regard, the NOD model provides an unambiguous demonstration that environmental factors indeed play a major role in regulation of genes determining the diabetic phenotype. The finding that immunostimulation is actually protective in prediabetic NOD mice suggests that there may be alternatives to the use of immunosuppressive agents in future intervention trials initiated during the prodromal phases of IDDM in humans. An interesting epidemiological observation has been that children of families in higher socio-economic strata in the Montreal area have a significantly higher incidence of IDDM, even when corrected for ethnic stratification [47]. Although no explanation for this observation is yet available, it is intriguing to speculate (in view of the increased diabetes risk to NOD mice reared in a very clean environment) that children from wealthier families may have been more "immunologically privileged" (i.e. more sequestered from common childhood illnesses such as flus and colds) than children from lower socio-economic strata.

Diabetogenic Catalysts in Natural Ingredient Diets

In rodents, dietary manipulations can produce alterations in immune function. An example would be the strong impairment of Mϕ function in BB rats fed essential fatty acid-free diets [28]. Similar reduction in autoimmune lupus in mice have been observed in response to reduced caloric intake [48]. Commercially-available closed formula (meaning that the components may vary without notice) natural ingredient diets for maintenance of laboratory rodents are quite variable in composition. Depending upon the country, the supplier, and the formula, a natural ingredient diet may contain any of the following vegetable products: milled corn, wheat, alfalfa, grass, soybean, barley, oats, wheat germ, beet pulp, and brewer's yeast. In addition, they contain animal protein (usually milk proteins and/or fish or animal liver meal), fat (usually corn or soybean oil), salts, and vitamins. The published literature indicates that, in suitably SPF environments, the commonly used natural ingredient diets all promote diabetogenesis in BB rats and NOD mice. The sharp reduction in diabetes incidence in BB rats maintained on semi-purified diet AIN-76 (all protein is supplied by casein, all carbohydrate by dextrin starch, and lipid by corn oil) has been discussed above. Studies with the NOD mouse have confirmed the presence of diabetogenic "catalysts" in natural ingredient diets. Elliott et al. observed that the 25% diabetes incidence obtained by feeding natural ingredient diet to a substrain of NOD maintained in New Zealand was reduced to 1% when the mice were maintained from weaning on Pregestimil, a hypoallergenic infant formula containing no vegetable meals and containing casein hydrolysate in place of casein [49]. Addition to Pregestimil of 10% (w/w) casein, but not 10% meat meal (present in the diet 86 New Zealand Stockfoods chow formulation) partially reconstituted diabetes incidence [22]. Finland has the highest incidence of IDDM in the world; it has been speculated that this may relate to the high per capita consumption of diary products [50]. Although cow's milk proteins in human diets are suspected as possible diabetogenic catalysts for initiation of autoimmune IDDM development, no compelling epidemiological evidence is available yet. Scott et al. suggested that the protective effects elicited by Pregestimil feeding may be attributable to the lack of intact proteins in this diet coupled with the lack of other diabetogenic triggers present in natural ingredient diets [22]. Semi-purified diets lack the bulk of vegetable matter (other than corn and coconut oils) found in natural ingredient grain-based diets. The possibility that metabolites of microbial or plant origin capable of modulating diabetogenic rate may be present in the food chain and may accumulate in the milk of ruminant animals will be discussed below.

Isolation of Diabetogenic Catalysts

Using females from our high diabetes incidence NOD/Lt colony at TJL, we have compared diabetes development in females fed from weaning an open formula (meaning constituents not subject to change without notice) natural ingredient-based chow diets [Old Guilford (OG) 96 and 96W] with females fed either semi-purified diet AIN-76 or Pregestimil. Diabetes incidence in our standard NOD/Lt colony maintained on our open formula non-purified OG 96W diet is currently 80%–90% in females and 40%–50% in males by 42 weeks of age. AIN-76 feeding from weaning failed to suppress insulitis development, yet significantly retarded the onset and the prevalence of diabetes. AIN-76 contains 20% of casein as a protein source; supplementation with 10% (w/w) skimmed milk proteins or 10% wheat gluten failed to accelerate diabetes development or to increase diabetes incidence [51]. Supplementation of AIN-76 with 10% brewer's yeast elicited a more rapid diabetes progression and higher prevalence than observed on AIN-76 alone but this addition was not as diabetogenic as a supplement of 25% OG 96 to the AIN-76 base diet [51]. It should be noted that there appear to be "hierarchies" of factors influencing the penetrance of the diabetes susceptibility genes. Whereas AIN-76 was found to be diabetes-retarding at TJL, where male incidence was no higher than 70%, the same diet was not capable of retarding diabetogenesis in the barrier-maintained NOD colony at the University of Alberta/Edmonton where a spontaneous diabetes incidence of 100% in both sexes was observed (Dr B. Singh, personal communication).

Pregestimil, containing no intact proteins, was effective in sustained blockage of both insulitis development and onset of the diabetic syndrome in NOD/Lt to 1 year of age, a time at which the incidence in the control group of females fed diet OG 96W had reached 95%. Insulitis was detected in the pancreas of one of three Pregestimil-fed females necropsied at 30 weeks of age and one of the nine Pregestimil-fed females that had remained diabetes-free developed diabetes at 54 weeks [51]. Studies are now in progress (Fig. 3.2) to establish whether pups to be weaned onto Pregestimil will be permanently protected from diabetes development if the breeding colony is maintained on Pregestimil (to eliminate the possibility that the nursing dams can adoptively transfer diabetogenic catalysts through the milk).

Fatty acid content of the diet is an important determinant of the spectrum of eicosanoids produced by metabolism of archidonic acid, including prostaglandins, thromboxanes and leukotrienes. These prostanoids in turn could lead to perturbations in immunological, as well as pancreatic islet secretory functions [25,26]. In contrast to fish oils rich in ω-3 long chain polyunsaturated fatty acids, linoleic-containing vegetable oils tend to increase cellular arachidonic acid content. It might be anticipated, therefore, that supplementation of a chow diet fed to NOD mice with long chain polyunsaturated fish oil might be more protective than with safflower oil. This was not the case [22]. We have found that at least some of the diabetogenic accelerators in the natural ingredient OG 96 are chloroform : methanol soluble, suggesting their lipoidal nature [51]. This was confirmed by the further extraction into heptane of the diabetogenic catalyst(s) isolated in the chloroform : methanol fraction [51]. Although the heptane fraction would contain corn oil (present in OG 96 at 5.8 g per 100 g) and its deriviatives, it seems unlikely that this is the source of the diabetogenic accelerator for several reasons. Corn oil is also the sole lipid source in the diabetes-retardant AIN-76 diet at 5 g per 100 g, and it is also present in the

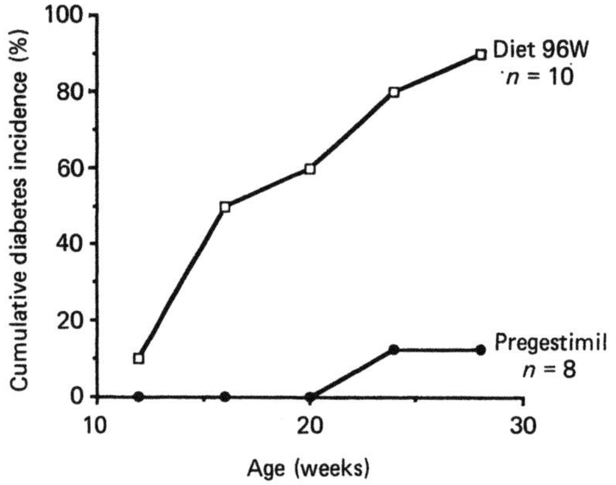

Fig. 3.2. Comparison of diabetes incidence in NOD/Lt females produced from breeders maintained on a complex natural ingredient diet (96W) versus those produced from breeders maintained on a semi-defined diet (Pregestimil). The females maintained on Pregestimil are clearly more diabetes-resistant than those fed the 96W natural ingredient diet.

strongly antidiabetogenic Pregestimil at higher concentration (10.5 g per 100 g) than in OG 96. Pregestimil also contains 7.7 g per 100 g of fractionated coconut oil, a more saturated oil than corn oil (the latter enriched in n-6 fatty acids, primarily linoleic acid). Increasing the fat content of a modified AIN-76 diet to 20% (5% corn oil plus 15% lard) failed to reconstitute accelerated diabetogenesis, as was produced by supplementing AIN-76 with 25% OG 96 meal. The biochemical nature of the lipoidal material in the chow diets is currently under investigation. Since factors in brewer's yeast also accelerated the rate of diabetogenesis when supplemented into AIN-76, the complex natural ingredients in a chow diet may contain a plethora of interactive substances capable of modifying diabetes susceptibility. Figure 3.2 illustrates the powerful effects of diet on control of diabetes penetrance.

Mechanism of Action of Diabetogenic Catalysts

Immunological modulation

Macrophages are central to pathogenesis in both the NOD and the BB models [27,52]. Not surprisingly, therefore, feeding an essential fatty acid deficient diet that prevented Mϕ migration into the islets and depleted basal concentrations of resident leukocytes in the islets, also markedly protected BB/W rats [28]. Thus, although essential fatty acids must be present in the diet for generation of functional diabetogenic effector cells, there is as yet no evidence linking diabetogenesis with either the content or degree of saturation of individual dietary lipids in BB rats [23] and NOD mice [22]. Children with newly diagnosed IDDM have increased levels of antibodies to cow's milk proteins [53]. Since bovine serum albumin (BSA) is strongly immunogenic in NOD mice [54], it might be argued that BSA shares some degree of homology with a β cell specific antigen such that an immune response to BSA elicits anti-islet autoimmunity. Indeed, rat antiserum against BSA binds in western blots to a 69,000 Mr protein produced in a rat insulinoma cell line [55]. In NOD/Lt mice, the presence or absence of an immune response to dietary BSA is not correlated to diabetogenesis. No anti-BSA antibodies are detectable by ELISA in serum from NOD/Lt mice maintained on OG 96W (our most diabetogenic diet) or Pregestimil (our most protective diet), whereas higher titres are found in serum from mice fed the moderately protective casein-enriched semi-purified diet [56]. Since digestion of casein produces biologically active peptides capable of stimulating Mϕ phagocytosis [57], it seems likely that protective immunostimulation rather than further impaired immunoregulation would result.

The grain and oil components of natural ingredient diets may be contaminated with bacteria, fungi, and/or their metabolites. Certain bacterial antigens, such as staphylococcal enterotoxin B, are considered to be "superantigens" in terms of ability to stimulate T lymphocytes and affect intrathymic tolerance development [58]. Orally administered antigens or microbial metabolites may be important in modification of a variety of immune responses, including activation of T-suppressor cells [59]. Fungal metabolites with well-known β-cytotoxic action have been described, the best known of which is SZ, produced by *Streptomyces achromogenes*. In view of the female gender bias of diabetes in NOD mice, another fungal metabolite with potential diabetogenic significance is zearalenone. Produced by the mould *Gibberella zeae* in roots of corn plants, this substance has weak oestrogenic properties [60]. In addition to SZ and zearalenone, fungal-derived alkaloid metabolites are known to concentrate in plants. Many of the agricultural "small grains" and many different grasses are parasitized by the ascomycetic fungus *Acremonium coenophialum*. Fields of tall fescue (*Festuca arundinacea* Schreb), the predominant cool season forage grass in the United States are commonly infected with *Acremonium coenophialum*. Under certain conditions, ergot alkaloids can be produced by this endophytic fungus that may attain concentrations in grass as high as 14 mg kg^{-1} in sheaths and 1.5 mg kg^{-1} in blades [61]. Ruminants are very efficient in extracting ergot alkaloids from plant tissue, and because of the vasoconstrictive activity of the alkaloids, they can develop toxic syndromes, including gangrene of the extremities [62]. Feeding CD-1 mice of both sexes either mouse chow containing 50% (w/w) endophyte-infected or uninfected grass seed [61] or rats of both sexes methanol-extracted alkaloids from *Acremonium coenophialum* (endophyte)-infected tall fescue seeds [63] severely depressed a wide variety of reproductive parameters. One of these parameters was lactation in females, and the physiological correlate of this was depressed serum prolactin levels [64,65].

The effects of prolactin on lactation and reproductive organs are well established; prolactin also has strong immunopotentiating properties and therefore may be essential for maintenance of normal immunoregulatory functions. Bromocryptine, an ergot alkaloid deriviative with dopamine type 2 agonist function produces hypoprolactinaemia in C3H/HeN male mice [66]. Continuously suppressing serum prolactin to less than 2 ng ml^{-1} by bromocryptine administration also suppressed Mϕ activation, a consequence of which was depressed mitogen stimulated T lymphocyte proliferation [66]. In addition, bromocryptine suppression of gamma-interferon production by T lymphocytes, the cytokine apparently necessary for Mϕ activation, was also implicated in the Mϕ dysfunction [66]. These immunological disturbances were reversed by prolactin administration. What is striking about these observations is that NOD/Lt mice are characterized by cytokine deficiencies which in turn are associated with dysfunctions in a Mϕ-T suppressor cell axis [34]. As reviewed above, the activation of T-suppressor cells in an SMLR is defective in NOD/Lt mice, but can be corrected by treatment in vivo with IL-2 [43]. The lipoidal dietary accelerators of diabetogenesis in OG 96 or OG 96W are potentially substances of microbial origin that either facilitate activation of precommitted T cytotoxic lymphocytes (CTL), or alternatively inactivate normal paracrine communication between self-MHC class II-presenting Mϕ and T suppressor–inducer cells.

Direct Effects on β Cells

Although the immune system and/or the endocrine system would seem to be the most logical targets of diet-derived regulators, the pancreatic islets themselves deserve serious consideration as targets of diabetogenic accelerators in the diet. SZ is but one of a host of β cytotoxins produced by fungi (e.g. cyclosporin inhibits mouse β cell function). SZ is a nitrosoamide with a glucose moiety which targets it to the β cell; once inside the cell, the nitrosoamide component is capable of decomposing to form highly reactive carbonium ions. These can alkylate various cellular components such as DNA and proteins, producing DNA strand breaks in the nucleus, and depletion of NAD and ATP levels in the cytoplasm [8]. In males of certain diabetes-susceptible inbred strains of mice, MSZ treatment induces expression of endogenous retroviral genomes in prenecrotic β cells, suggesting that the environmental insult can elicit the expression of neoantigenic products that would attract the attention of the immune surveillance system [8]. A correlation between the initiation of severe insulitis and induction of these endogenous retrovirus-like particles in NOD males has recently been reported [67]. Cumulative incidence of diabetes in 24-week-old NOD males in the colony of these investigators was only about 10%, but diabetes could be induced rapidly by high doses of cyclophosphamide (150 mg kg^{-1} body weight twice every 3 days). Induction of high levels of an aberrant intracisternal retroviral particle was observed in the prenecrotic β cells of these males. Insulin administered to prediabetic BB rats [68] and NOD mice [69] prevents diabetes onset. It has been suggested that this therapy, by lowering plasma glucose, metabolically "rests" pancreatic β cells such that they express lower levels of autoantigens, including retroviral products (see above discussion). β cells are thought to be particularly sensitive to free radical induced damage, such that diets rich in substances stimulating inflammatory cell activity and generation of free radicals would be expected to be diabetogenic whereas diets containing agents that control free radicals or their effects (vitamin E, nicotinamide) might be expected to be protective. As discussed above, xenobiotic substances in diet that function to activate Mϕ also protect against diabetogenesis in NOD mice. Regarding the presence or absence of factors in diets that may control free radicals or their effects, 1% nicotinamide supplemented in the drinking water did protect chow fed NOD mice from diabetes development, whereas 1 mg ml^{-1} vitamin E was not protective [22]. Dimethyl sulphoxide when administered in drinking water was expected to attenuate the incidence of diabetes in chow fed males; instead it increased incidence from 21% to 79% [70]. Regarding the question of molecules in the diet capable of mediating direct β cell cytotoxicity, nitrite addition is a common method of curing food and stabilizing certain beverages such as wine. In Iceland, IDDM developed in some male offspring of mothers who, during pregnancy, ingested cured mutton containing high levels of *N*-nitroso compounds [9]. When this mutton or the *N*-nitroso compounds identified in the meat was fed to CD-1 male mice (which are highly sensitive to MSZ-induced diabetes) the mice developed diabetes [9].

Summary and Conclusions

This review has addressed the question of whether environmental factors represent important or necessary "triggers" required to initiate the onset of IDDM, using knowledge and insights gained from the BB rat and the NOD mouse. Both rodent models are characterized by defective immunoregulatory networks; the surprising finding has been that general immunostimulation by environmental agents, or dietary manipulations, are both capable of retarding or preventing the onset of diabetic hyperglycaemia in each model. This discovery suggests new possibilities for circumventing the onset of diabetes in humans identified as at high risk to develop IDDM. Instead of the immunosuppressive therapies contemplated now that may leave the individual severely immunocompromised, appropriate levels of immunostimulation may be devised that activate a more normal immunoregulatory network capable of suppressing the activation of autoreactive immune effector cells. At the same time, if the dietary catalysts that accelerate onset of diabetes in NOD mice and BB rats can be identified, elimination or reduction of these elements from human foodstuffs may also be beneficial.

Acknowledgements. The writing of this review was supported by a grant from The Juvenile Diabetes Foundation, International, and by NIH grants DK 27722 and 36175. In addition to thanking all of his scientific colleagues for their contributions to this review, the author would like to acknowledge the "grass roots" support of all the people of The Juvenile Diabetes Foundation, International. These people are labouring to support the international scientific research efforts that will provide answers to the many unanswered questions raised in this review.

References

1. Eisenbarth GS (1986) Type 1 diabetes mellitus: a chronic autoimmune disease. N Engl J Med 314:1360–1368
2. Sibley RK, Sutherland DE, Goetz F, Michael AF (1985) Recurrent diabetes in the pancreas iso- and allograft. A light and electron micrographic and immunohistochemical analysis of four cases. Lab Invest 53:132–144
3. Sheehy MJ, Scharf SJ, Rowe JR, Neme de Gimenez MH, Meske LM, Ehrlich HA, Nepom BS (1989) A disease susceptible HLA haplotype is best defined by a combination of HLA-DR and -DQ alleles. J Clin Invest 83:830–835
4. Todd JA, Acha-Orbea H, Bell JI, Chao N, Fronek Z, Jacob CO, McDermott M, Sinha AA, Timmerman L, Steinman L, McDevitt HO (1988) A molecular basis for MHC class II-associated autoimmunity. Science 240:1003–1009
5. Barnett AH, Eff C, Leslie RDG, Pyke DA (1981) Diabetes in identical twins: A study of 200 pairs. Diabetologia 20:87–93
6. Scott FW, Daneman D, Martin JM (1988) Evidence for a critical role of diet in the development of insulin-dependent diabetes mellitus. Diabetes Res 7:153–157
7. Rayfield EJ, Yoon JW (1981) Role of viruses in diabetes. In: Cooperstein SJ, Watkins D (eds) The islets of Langerhans: biochemistry, physiology, and pathology. Academic Press, New York pp 428–451
8. Wilson GL, Leiter DH (1990) Streptozotocin interactions with pancreatic β cells and the induction of insulin dependent diabetes. In: Dyrberg T (ed) Current topics in microbiology and immunobiology. Springer, Berlin 156:27–54
9. Helgason T, Ewen SWB, Ross IS (1982) Diabetes produced in mice by smoked/cured mutton. Lancet ii:1017–1022
10. Eisenbarth GS (1986) Genes, generator of diversity, glycoconjugates, and autoimmune β-cell insufficiency in type I diabetes: Diabetes 36:355–364
11. Tonegawa S (1988) Somatic generation of immune diversity. In Vitro Cell Devel Biol 24:253–265
12. Acha-Orbea H, Mitchell DJ, Timmerman L, Wraith DC, Tausch GS, Waldor MK, Zamvil SS, McDevitt HO, Steinman S (1988) Limited heterogeneity of T cell receptors from lymphocytes mediating autoimmune encephalomyelitis allows specific immune intervention. Cell 54:263–273
13. Mordes JP, Desemone J, Rossini AA (1987) The BB rat. Diabetes/Metabolism Rev 3:725–750
14. Guberski DL, Butler L, Kastern W, Like AA (1989) Genetic studies in inbred BB/Wor rats.

Analysis of progeny produced by crossing lymphopenic diabetes-prone rats with nonlymphopenic diabetic rats. Diabetes 38:887–893
15. Woda BA, Handler ES, Padden C, Greiner DL, et al. (1987) Anti-asialo GM1 (AGMI) prevents diabetes in diabetes prone (DP) but not RT6.1 depleted diabetes resistant (DR) BioBreeding/Wor rats. Diabetes 36 (Suppl 1):39A
16. Colle E, Guttman RD, Fuks A, Seemayer TA, Prud'homme GJ (1986) Genetics of the spontaneous diabetic syndrome: interaction of MHC and non-MHC associated factors. Molec Biol Med 3:13–23
17. Butler L, Guberski DL, Like AA (1983) Genetic analysis of the BB/W diabetic rat. Can J Genet Cytol 25:7–15
18. Rossini AA, Williams RM, Mordes JP, Appel MC (1979) Spontaneous diabetes in the gnotobiotic BB/W rat. Diabetes 28:1031–1032
19. Rossini AA, Mordes JP, Gallina DL, Like AA (1983) Hormonal and environmental factors in the pathogenesis of BB rat diabetes. Metabolism 32 (Suppl 1):33–36
20. Ahmed SA, Dauphinee MJ, Talal N (1985) Effects of short term administration of sex hormones on normal and autoimmune mice. J Immunol. 134:204–210
21. Good RA (1981) Nuitrition and immunity. J Clin Immunol 1:3–10
22. Scott FW, Elliott RB, Kolb H (1989) Diet and autoimmunity: prospects for prevention of type 1 diabetes. Diabetes, Nutrition Metabolism 2:61–73
23. Issa-Chergui B, Guttmann RD, Seemayer TA, Kelley VE, Colle E (1988) The effect of diet on the spontaneous insulin dependent diabetic syndrome in the rat. Diabetes Res 9:81–86
24. Kelley VE, Ferretti A, Izui S, Strom TB (1985) A fish oil diet rich in eicosapentaenoic acid reduces cyclooxygenase metabolites and suppresses lupus in MRL-Ipr mice. J Immunol 134:1914–1919
25. Robertson RP (1988) Eicosanoids as pluripotential modulators of pancreatic islet function. Diabetes 37:367–370
26. Thomas IK, Erickson V (1985) Dietary fatty acid modulation of murine T-cell responses in vivo. J Nutr 115:1528–1534
27. Haneberg H, Kolb-Bachofen V, Kantwert-Funke G, Kolb H (1989) Macrophage infiltration precedes and is a prerequisite for lymphyocytic insulitis in pancreatic islets of prediabetic BB rats. Diabetologia 32:126–134
28. Schreiner G, Rossini A, Mordes J, Handler E, Lacy P, Wright J, Lefkowith J (1988) Essential fatty acid deficiency inhibits the development of diabetes in rats. Diabetes 37 (Suppl 1):80A
29. Schwimmbeck PL, Dyrberg T, Oldstone MBA (1988) Abrogation of diabetes in BB rats by acute virus infection J Immunol 140:3394–3400
30. Satoh J, Shintani S, Oya K, Nobunaga T et al. (1987) Treatment with a streptococcal preparation (OK-432) suppresses anti-autoimmunity and prevents diabetes in BB rats. Diabetes 36 (Suppl 1):66A
31. Makino S, Kunimoto K, Muraoka Y, Mizushima Y, Katagiri K, Tochino Y (1980) Breeding of a non-obese, diabetic strain of mice. Exp Anim 29:1–8
32. Prochazka M, Leiter EH, Serreze DV, Coleman DL (1987) Three recessive loci required for insulin-dependent diabetes in NOD mice. Science 237:286–289
33. Serreze DV, Leiter EH, Worthen SM, Shultz LD (1988) NOD marrow stem cells adoptively transfer diabetes to resistant (NOD × NON) F1 mice. Diabetes 37:252–255
34. Serreze DV, Leiter EH (1988) Defective activation of T suppressor cell function in nonobese diabetic mice. Potential relation to cytokine deficiencies. J Immunol 140:3801–3807
35. Serreze DV, Hamaguchi K, Leiter EH (1988) Defective immunoregulatory signaling in NOD mice. Diabetes 37 (Suppl 1):98A
36. Hatamori N, Shii K, Sakamoto T, Yaso S, Hari J, Aoyama N, Doi K, Taniguchi H (1988) Impaired expression of high affinity interleukin-2 receptors on activated T lymphocytes from NOD mice. Diabetes 37 (Suppl 1):388A
37. Suzuki T, Yamada T, Takao T, Fujimura T, Kawamura E, Shimizu ZM, Yamashita R, Nomoto K (1987) Diabetogenic effects of lymphocyte transfusion on the NOD or NOD nude mouse. In: Rygaard J, Graem NBN, Sprang-Thomsen M (eds) Immune-deficient animals in biomedical research. Karger, Basel, pp 112–116
38. Leiter EH, Wilson GL (1988) Viral interactions with pancreatic β cells. In: Lefebvre PJ, Pipeleers D (eds) The pathology of the endocrine pancreas in diabetes. Springer, Berlin, pp 85–105
39. Oldstone MBA (1988) Prevention of type 1 diabetes in Nonobese Diabetic Mice by virus infection. Science 23:500–502
40. Leiter EH, Le PH, Prochazka M, Worthen SM, Huppi K (1988) Genetic and environmental

control of diabetes induction by multi-dose streptozotocin in two BALB/c substrains. Diabetes Res 9:5–10
41. Pukel C, Baquerizo H, Rabinovitch A (1988) Destruction of rat islet cell monolayers by cytokines. Synergistic interactions of interferon-g, tumor necrosis factor, lymphotoxin, and interleukin 1. Diabetes 37:133–136
42. Blay RA, Bigley NJ, Giron DJ (1985) A murine model of insulin-dependent diabetes mellitus resulting from the cumulative effects of the non diabetogenic strain of encephalomyocarditis virus and a single low dose of streptozotocin. Diabetes 34:1288–1292
43. Serreze DV, Hamaguchi K, Leiter EH (1990) Immunostimulation circumvents diabetes in NOD/Lt mice. J Autoimmunity 2:759–776
44. Sadelain MWJ, Lauzon J, Singh B (1989) Early adjuvant immunotherapy in Non-Obese Diabetic (NOD) mice prevents diabetes and increases endogenous natural suppressor activity. Diabetes 38 (suppl 2):71A
45. Toyota T, Satoh J, Oya K, Shintani S Okano T (1986) Streptococcal preparation (OK-432) inhibits development of type 1 diabetes in NOD mice. Diabetes 32:496–499
46. Bonavida B, Jewett A (1989) Activation of human peripheral blood-derived monocytes by OK-432 (*Streptococcus pyogenes*): augmented cytotoxicity and secretion of TNF and synergy with rIFN-γ. Cell Immunol 123:373–383
47. Colle E, Siemiatycki J, West R et al. (1981) Incidence of juvenile onset diabetes in Montreal: demonstration of ethnic differences and socioeconomic class differences. J Chron Dis 34:611–616
48. Fernandes G, Friend P, Yunis EJ, Good RA (1978) Influence of dietary restriction on immunologic function and renal disease in (NZB × NZW)F1 mice. Proc Natl Acad Sci USA 75:1500–1504
49. Elliott RB, Reddy SN, Bibby NJ, Kida K (1988) Dietary prevention of diabetes in the non-obese diabetic mouse. Diabetologia 31:62–64
50. LaPorte RE, Tajima N, Akerbloom HK et al. (1985) Geographic differences in the risk of insulin-dependent diabetes mellitus; the importance of registries. Diabetes Care 8 (Suppl 1):101–107
51. Coleman DL, Kuzava JE, Leiter EH (1990) Effect of diet on the incidence of diabetes in non-obese diabetic (NOD) mice. Diabetes 39:432–436
52. Lee K-U, Amano D, Yoon J-W (1988) Evidence for the initial involvement of macrophage in development of insulitis in NOD mice. Diabetes 37:989–991
53. Savilhati E, Akerbloom H, Taino V-M, Koskimies S (1988) Children with newly diagnosed insulin dependent diabetes mellitus have increased levels of cow's milk antibodies. Diabetes Res 8:137–140
54. Beppu H, Winter WE, Atkinson MA, Maclaren NK, Fujita K, Takahashi H (1987) Bovine albumin antibodies in NOD mice. Diabetes Res 6:67–69
55. Glerum M, Robinson BH, Martin JM (1989) Could bovine serum albumin be the initiating antigen ultimately responsible for the development of insulin dependent diabetes? Diabetes Res 10:103–107
56. Serreze DV, Leiter EH, Kuff EL, Jardieu P, Ishizaka K (1988c) Molecular mimicry between insulin and retroviral antigen p73. Development of cross reactive autoantibodies in sera of NOD and C57BL/KsJ *db/db* mice. Diabetes 37:351–358
57. Migliore-Samour D, Jolles P (1988) Casein, a prohormone with an immunomodulating role for the newborn? Experientia 44:188–193
58. White J, Herman A, Pullen AM, Kubo R, Kappler JW, Marrack P (1989) The Vβ-specific superantigen staphylococcal enterotoxin B; stimulation of mature T cells and clonal deletion in neonatal mice. Cell 56:27–35
59. Mowat AM (1987) The regulation of immune responses to dietary protein antigens. Immunol. Today 8:93–98
60. Mastri C, Mistry P, Lucier GW (1985) In vitro oestrogenicity and binding characteristics of α-zearalanol (P-1496) to different classes of oestrogen binding proteins in rat liver. J Steroid Biochem 23:279–289
61. Zavos PM, Varney DR, Siegel MR, Hemken RW, Jackson JA, Bush LP (1987) Effects of feeding endophyte-infected tall fescue seed on the reproductive performance in male and female CD-1 mice by combination crosses. Theriogenology 27:541–548
62. Lyons PC, Plattner RD, Bacon CW (1986) Occurrence of peptide and clavine ergot alkaloids in tall fescue grass. Science 232:487–489
63. Zavos PM, Varney DR, Bush LP, Hemken RW, Jackson JA, Siegal MR (1988) Reproductive responses in male rats fed extracts of fescue seed infected by the endophytic fungus *Acremonium coenophialum*. Drug Chem Toxicol 11:113–133
64. Jackson JA, Hemken RW, Bush LP, Boling JA, Siegel MR, Zavos PM (1987) Physiological

responses in rats fed extracts of endophyte infected tall fescue seed. Drug Chem Toxicol 10:369–379
65. Jackson JA, Varney DR, Hemken RW, Siegel MR, Bush LP, Harmon RJ, Wise S, Gardner P (1986) Use of serum prolactin in rats as a determinant in detecting endophyte infected tall fescue seed. Drug Chem Toxicol 9:33–39
66. Bernton EW, Meltzer MS, Holaday JW (1988) Suppression of macrophage activation and T-lymphocyte function in hypoprolactinemic mice. Science 239:401–404
67. Suenaga K, Yoon J-W (1988) Association of β-cell-specific expression of endogenous retrovirus with development of insulitis and diabetes in NOD mouse. Diabetes 37:1722–1726
68. Gotfredsen CF, Buschard K, Frandsen E (1985) Reduction of diabetes incidence of BB rats by early prophylactic insulin treatment of diabetes-prone animals. Diabetologia 28:933–935
69. Atkinson MA, Maclaren NK, Luchetta R, Burr I (1989) Prophylactic insulin therapy prevents insulitis and insulin dependent diabetes (IDD) in NOD mice. Diabetes 38 (Suppl 2):87A
70. Klandorf H, Chirra AR, DeGruccio A, Girman DJ (1989) Dimethyl sulfoxide modulation of diabetes onset in NOD mice. Diabetes 38:194–197

Chapter 4

The Role of the Microflora in Graft-versus-Host Disease After Allogeneic Bone Marrow Transplantation

D. W. van Bekkum and P. J. Heidt

Introduction

Currently, allogeneic bone marrow transplantation (BMT) is an accepted treatment for many fatal diseases of the haemopoietic system, among them severe aplastic anaemia [1] and leukaemia [2]. Furthermore, patients suffering from fatal hereditary diseases that are associated with a dysfunction of the lymphoid system, such as severe combined immunodeficiency [3], and patients with inherited severe metabolic disorders [4] are being treated with bone marrow grafts. In total about 2500 patients are transplanted annually world-wide in 160 BMT centres.

One of the major complications of allogeneic BMT is graft-versus-host disease (GvHD), which is caused by donor type lymphocytes which react against the recipients tissues. According to a recent evaluation of data from 2036 recipients of HLA identical sibling bone marrow transplants reported to the International Bone Marrow Transplant Registry, moderate to severe GvHD occurred in about 45% of these patients. In 48% of them, GvHD was related to their death [5]. The severity of GvHD is influenced by several factors, which include the degree of immunogenic disparity [6], the number of cells grafted [7], the number of T lymphocytes present in the graft [7,8], the donor's sex [5] and the age of the recipient [5].

GvHD is a more frequent and more severe complication of allogeneic BMT in primates than in rodents, because of an approximately tenfold higher proportion of T lymphocytes in the bone marrow of primates. Severe acute GvHD starting before day 14 after BMT is hardly ever seen in rodents but is normal in the matched and mismatched rhesus monkey combinations. The GvHD of mice can be aggravated reproducibly and in a graded fashion by adding increasing numbers of lymph node or spleen cells to the bone marrow graft. Conversely, GvHD in primates and rodents can be prevented by removing or inactivating the T lymphocytes in the graft prior to transplantation.

Microflora and Graft-versus-Host Disease in Mice

Another important factor influencing GvHD is the recipient's gastrointestinal microflora. This was originally observed in gnotobiotic mice. In contrast to the bone marrow of primates, rodent bone marrow contains a low proportion of immunocompetent T lymphocytes. As a consequence, its GvHD-inducing potential is comparatively low [9]. Infusion of 10^7 H-2 incompatible bone marrow cells into lethally irradiated (9.0 Gy X-rays) conventional mice results in a late onset type GvHD which does not give rise to symptoms until about 3 weeks after BMT. This disease kills the majority of the recipients during the next 2 months but those that survive for more than 3 months seem to have recovered [9,10]. This type of GvHD is called delayed GvHD to distinguish it from the acute GvHD which can be induced in mice by supplementing the allogeneic bone marrow graft with donor derived spleen or lymph node cells. Mortality attributable to delayed GvHD can be completely prevented if the recipients are germ-free mice [10–12], or when they are conventional animals that have been subjected to complete [12,13] or selective [10] gastrointestinal decontamination (GID) by means of orally administered nonabsorbable antibiotics prior to transplantation. In these experiments, selective decontamination resulted in animals harbouring only a strict anaerobic microflora.

If the bone marrow graft of conventional mice is supplemented with 10^7 spleen cells so as to provide a graft composition that is, like primate marrow, rich in immunocompetent T lymphocytes, the recipients suffer from an early onset GvHD that takes a fatal course within 3 weeks of transplantation. The absence of a gastrointestinal microflora in the recipients (i.e. germ-free or completely decontaminated animals) delays this mortality by two weeks, but does not prevent it [10].

Jones et al. [11] showed that conventionalization of germ-free or completely decontaminated recipients of allogeneic H-2 mismatched bone marrow at 150 and 180 days after BMT caused their death within 4 weeks. It is conceivable that this mortality was the result of infection caused by uncontrolled colonization of the gastrointestinal tract, since Heit and colleagues [13] showed that reconventionalization of the completely decontaminated chimeras, starting 175 days after BMT, did not produce mortality. We investigated the time period after BMT during which the recipients had to be maintained germ-free or decontaminated in order not to lose the mitigating effect of the gnotobiotic state on delayed GvHD. It became clear that reconventionalization at day 40 or later after BMT, did not influence the beneficial effect of the decontaminated or germ-free state; also over 90% of these chimeras survived after reconventionalization without showing any signs of GvHd [10]. We also studied the effect of earlier reconventionalization after BMT (i.e. days 8, 20 and 26 after BMT). Reconventionalization at days 8 and 20 resulted in a mortality pattern after BMT identical to that of conventional recipients of allogeneic BMT. Reconventionalization at day 26 after BMT gave a protective effect in about 50% of the recipients, the remaining animals survived for more than 200 days after BMT [14]. The results of these experiments show that the severity of GvHD is determined by the presence or absence of (some constituents) of the recipient's microflora. These findings suggested that not only histoincompatibility is determining the occurrence and severity of GvHD, but that microflora-related factors also are of major importance. The hypothesis was that donor type lymphocytes can be stimulated or activated by antigens of bacteria from the gastrointestinal tract which cross-react with antigens present on the recipient's epithelial tissues, these being the main targets in GvHD. A key role in this phenomenon was attributed to the Enterobacteriaceae, which were absent in the germ-free, selectively or completely decontaminated, and CRF recipients of allogeneic mismatched bone marrow in which GvHD was mitigated, while this group of bacteria was always present in the conventional recipients, of which 90% died from severe GvHD.

Fetal Gut Implants

The hypothesis that GvHD is the result of T cell stimulation, or of activation by cross-reactive antigens present on enteric bacteria and the recipient's epithelium was confirmed by experiments performed by van Bekkum and Knaan [15]. In (CBA × C57BL) F1 hybrid mice they implanted CBA and F1 fetal gut fragments. Because the fetal implants remain germ-free, this technique allows the study of intestinal GvHD in sterile and conventional conditions in the same host. By the use of parent into F1 hybrid combinations the reactions to syngeneic and allogeneic intestines can also be investigated in the same mouse. After the implants were established (about 15 to 30 days after implantation), the mice were irradiated (9.0 Gy X-rays) and transplanted with 2×10^6 CBA bone marrow plus 2×10^6 CBA spleen cells, so that they developed acute GvHD. This was done in conventional as well as in completely decontaminated carriers of fetal gut implants. As a control, conventional and completely decontaminated F1 carriers of fetal gut implants were transplanted with similar amounts of F1 bone marrow and spleen cells. Scoring of GvHD in the different groups was done by counting the numbers of degenerated and intact crypts in sections of the implanted F1 or CBA guts. The results are summarized in Table 4.1. After transplantation of CBA bone marrow and spleen cells in conventional F1 mice, the damage in the F1 fetal gut implant was twice as great as in F1 fetal gut implants of the decontaminated recipients. In conventional recipients of CBA bone marrow and spleen cells, the CBA fetal gut implants which were not in direct contact with any microflora showed a significant score of GvHD lesions, while in decontaminated chimeras the CBA implant showed no histological lesions. None of the implants of the control animals, which had been treated with syngeneic cells, showed histopathological lesions characteristic for GvHD. It was concluded that the presence of a microflora at a distant site is capable of magnifying GvHD lesions in the germ-free F1 gut implant, and is even capable of inducing donor type immune cells to cause GvHD lesions in the germ-free CBA implant which is syngeneic to these cells.

Table 4.1. Crypt damage (%) in fetal gut implants of conventional and completely decontaminated mice

Chimera (donor → recipient)	Fetal gut score: % damaged crypts	
	F1	CBA
CBA → F1 (conventional)	47	23
CBA → F1 (decontaminated)	23	3
F1 → F1 (conventional)	<3	<3
F1 → F1 (decontaminated)	<3	<3

The above summarized observations all clearly indicate an important role for the microflora in the development and severity of GvHD after allogeneic bone marrow transplantation. More research was carried out to reveal the mechanism involved, while the possible clinical relevance of such observations was investigated in a preclinical model.

Preclinical Studies in Rhesus Monkeys

When no HLA identical sibling donor is available, bone marrow from an HLA mismatched donor can be used for transplantation. However, the use of mismatched donor bone marrow increases the risk of GvHD. In rhesus monkeys it was shown that there is a relation between the degree of immunogenetic disparity between donor and recipient and the incidence of lethal GvHD after allogeneic BMT. As a consequence, T lymphocyte depletion of bone marrow grafts has to be more efficient when immunogenetic disparity increases, in order to prevent lethal GvHD [16].

Experiments in mice have shown that GID mitigates GvHD after transplantation of bone marrow from an allogeneic, H-2 incompatible donor. This effect can only be observed with grafts containing low numbers of lymphocytes [17]. Accordingly, to evaluate this procedure in a preclinical model, rhesus monkeys grafted with T cell depleted, mismatched related and unrelated bone marrow were used [18, pp 74–93]. The recipients underwent either complete decontamination or selective decontamination, both performed under strict reverse isolation conditions.

With regard to the incidence and severity of GvHD, these gnotobiotic animals were compared with a historical group of recipients which were comparable to our treatment groups with regard to the conditioning and the bone marrow grafts, but which were housed under "clean conventional" conditions [19]. A total of 19 monkeys was completely decontaminated, while selective decontamination was performed in 6 monkeys.

GID was considered successful when in the period from 7 days before until 40 days after BMT the target micro-organisms could not be isolated from more than three consecutive faecal samples. These target organisms were all bacteria, yeasts and fungi in the case of complete GID and all facultative anaerobic Gram-negative bacteria, *Staphylococcus aureus*, and yeasts and fungi in the case of selective GID.

After the animals were conditioned for BMT with total body irradiation (TBI), which was performed under gnotobiotic conditions, they were transplanted with stem cell enriched, lymphocyte depleted bone marrow grafts, obtained by using a discontinuous albumin density gradient as designed by Dicke and van Bekkum [20]. Retrospective studies have shown that this procedure eliminates approximately 90% of the T lymphocytes from the monkeys' bone marrow. The donors in this study were either unrelated-mismatched (n=12), unrelated-A/B matched (n=6), or related-D/DR matched (n=7). The cell dose was 5×10^7 per kg body weight.

As can be seen in Table 4.2, complete GID was successful in 12 of the 19 monkeys, while selective GID was successful in 4 of the 6 monkeys. GvHD could be studied in 16 of the 25 transplanted animals; of the remaining animals, four did not establish a take of the donor bone marrow, while five monkeys were not evaluable because of early death caused by other complications, mostly severe electrolyte imbalance. The incidence of lethal GvHD in the different donor/recipient combinations and their distribution over the different gnotobiotic groups of recipients is given in Table 4.3, while the extremes of survival times of the same groups of monkeys are given in Table 4.4.

Table 4.2. Success of complete and selective gastrointestinal decontamination of rhesus monkeys

Type of decontamination	Number of animals	
	Total	Successfully decontaminated
Complete	19	12 (63%)
Selective	6	4 (67%)

Table 4.3. Incidence of lethal GvHD in the different gnotobiotic groups of donor/recipient combinations

Donor/recipient combination			Gnotobiotic state		
RhLA A/B	D/DR	Family relationship	Complete GID	Selective GID	Clean conventional
=	≠	None	0/4[a]	–	5/9[b]
≠	=	Sibling	0/2[c]	0/1[d]	4/5[e]
≠	≠	None	6/8[f]	1/1[g]	5/5[h]

Chi-square test: $(a+c+d) - (b+e)$: $p<0.05$
$(f+g) - h$: $p>0.05$ (n.s.)

The different groups of decontaminated recipients with an established take of the bone marrow graft are too small to make any comparison between the two types of GID with regard to GvHD. Apart from this, it was found that engraftment of stem cell concentrates from RhLA-A/B matched, -D/DR mis-

Table 4.4. Survival time (extremes) of the different gnotobiotic groups of donor/recipient combinations

Donor/recipient combination			Gnotobiotic state		
RhLA A/B	D/DR	Family relationship	Complete GID	Selective GID	Clean conventional (days)
=	≠	None	60 days to >10 years	–	23–269
≠	=	Sibling	46–253 days	82	19–66
≠	≠	None	19 days to >1 year	25	17–39

matched unrelated donors occurred in four of the six completely decontaminated recipients, of which none developed lethal GvHD. The survival times of these animals were 60, 80, >3230 and >3794 days (median: 1655 days), the two latter animals still living and well 9 and 10 years after BMT. Similarly, completely decontaminated recipients of stem cell concentrates of RhLA-A/B mismatched, -D/DR matched sibling donors who had an established take (three out of seven animals) did not develop lethal GvHD (in only one animal transient GvHD of the skin was observed), resulting in prolonged survival. The group of decontaminated animals that were identical either for A/B or D/DR, developed significantly less (zero out of seven animals) GvHD than did their clean conventional (nine out of fourteen) counterparts ($p<0.05$; Table 4.3).

No favourable influence of the decontaminated state of recipients of T lymphocyte depleted stem cell concentrates of unrelated, completely mismatched donor bone marrow was observed. Of the nine out of twelve recipients who had an established take of the bone marrow graft, seven developed lethal GvHD. This is not significantly different from the 12 clean conventional animals, which all developed lethal GvHD (Table 4.3). The failure of gnotobiotic measures to prevent lethal GvHD in this group of recipients can be ascribed to the relatively large number of residual T lymphocytes in the grafts which were, as will be discussed later, only one log depleted. In this mismatched donor/recipient combination, in contrast to partially matched donor/recipient combinations, residual T lymphocytes are capable of evoking a GvHD which apparently cannot be mitigated by these gnotobiotic measures. This hypothesis is based on our observations in mice, which showed that a beneficial effect of the gnotobiotic state of the recipient of allogeneic H-2 mismatched bone marrow was only observed when relatively low numbers of spleen cells were added to the graft [21]. The survival times of the nine monkeys, all of which showed full engraftment, varied between 19 and 48 days after BMT (median: 24 days).

The method used for T lymphocyte depletion (discontinuous albumin density gradient centrifugation) resulted in the elimination of approximately 90% of the T lymphocytes. Application of one of the more recently developed techniques (E-rosetting techniques or the use of Campath monoclonal antibodies) will result in the elimination of about 99% of T lymphocytes, while addition of both techniques to the discontinuous albumin density gradient centrifugation technique finally results in the elimination of up to 99.90% to 99.99% of the T lymphocytes from the bone marrow graft. In a limited number of monkeys,

Wagemaker has demonstrated that GvHD can be completely prevented by these more efficient depletion procedures [16].

Clinical Bone Marrow Transplantation

In man, the effect of GID on GvHD has been the cause of controversy for a long time: several studies reported a reduction of the incidence of acute GvHD after allogeneic BMT [22–24]. However, such an effect was not observed in other studies [25–27]. Recently, Vossen and Heidt reported on a retrospective evaluation of the efficacy of GID in a protective environment for the prevention of GvHD in 65 children and adolescents, grafted consecutively for either severe bone marrow failure (n=29) or leukaemia (n=36) [28]. All patients received full bone marrow grafts from HLA genotypically identical siblings, following the usual pre-treatment regimens.

The patients were maintained in a strict protective environment, i.e. in laminar downflow isolators [29], using aseptic nursing techniques and sterilization of food, beverages and all other items brought into the isolator.

Forty-four recipients received high doses of nonabsorbable antimicrobial drugs for complete GID, and 21 underwent selective GID with either absorbable or low doses of nonabsorbable antimicrobial drugs. Decontamination started a week or more before the date of BMT and was given for a total of at least 40 days after BMT, according to experimental data [14]. Both complete and selective GID were considered successful when in the period from 7 days before until 40 days after BMT the target micro-organisms could not be isolated from more than two consecutive faecal samples. GvHD was diagnosed by clinical symptoms. The severity of acute GvHD was graded according to Thomas et al. [30].

According to the criteria used, GID was successful in 12 out of 44 children with complete GID (27%) and in 17 out of 21 children with selective GID (81%). The effect of GID on acute GvHD could be evaluated in 58 and 48 patients, respectively. Seven children were not evaluable for acute GvHD: one was grafted with bone marrow from a syngeneic donor, four patients died before 16 days after BMT, and two patients had no engraftment.

Table 4.5 summarizes the data on acute GvHD. None of the 11 successfully completely decontaminated recipients developed ≥ grade II acute GvHD. In the group of 29 children in which complete GID was unsuccessful, less acute GvHD was observed than in the group of 14 patients who were selectively decontaminated with success, i.e. 24% versus 43%, respectively, although this difference was not statistically significant. The group of completely decontaminated patients as a whole did significantly better with respect to acute GvHD than the selectively decontaminated groups, whether successful or not [(a+b) versus (c+d), c, and d; see Table 4.5].

From Table 4.5 it is apparent that complete GID is superior to selective GID in preventing ≥ grade II acute GvHD. In the latter treatment group the percentage of patients developing this disease was 50%, which is not significantly different from the incidence (45%) in 2036 recipients of HLA identical sibling bone marrow transplants analysed by the International Bone Marrow Transplant

Table 4.5. Effect of the success of gastrointestinal decontamination on the occurrence of ≥grade II acute graft-versus-host disease in paediatric patients

Number of patients	Type of GID	Success of GID		Number of patients with ≥grade II acute GvHD
40	Complete	Yes:	11	0[a]
		No:	29	7[b]
18	Selective	Yes:	14	6[c]
		No:	4	3[d]

Significant differences: $(a+b) - (c+d)$: $p<0.01$
$a - c$: $p<0.05$
$a - (c+d)$: $p<0.05$
$(a+b) - c$: $p<0.05$
$(a+b) - d$: $p<0.005$
All other differences: $p>0.05$ (n.s.)

Registry [5]. It is of interest that of the successfully selectively decontaminated patients in the present study, a comparable percentage (43%) developed this disease (Table 4.5).

The Underlying Mechanism

To study the mechanism underlying the influence of the gastrointestinal microflora on GvHD, H-2 different donor and recipient mice with a specific pathogen-free (SPF) and a conventional microflora were employed [18]. For this purpose a conventional microflora was imported from the M.D. Anderson Cancer Institute, Houston, TX, USA, called "Houston flora" (HF). Germ-free C3H/Law breeding pairs were associated with this flora. HF C57BL/Rij mice were obtained by foster nursing Caesarian-derived C57BL/Rij newborns by HF C3H/Law mothers. The HF C57BL/Rij animals obtained in this way were bred to produce the experimental animals. The mice were kept in plastic film isolators to prevent undue association with any other micro-organisms. In addition, the effect of complete and selective GID of HF recipients on GvHD was studied (Heidt et al. unpublished work).

The bone marrow recipients for the first series of experiments were 12- to 20-week-old C3H/Law ($H\text{-}2^k$) male and female mice. The donors of the bone marrow grafts were 12- to 20-week-old C57BL/Rij ($H\text{-}2^b$) male or female mice. In all experiments, the sex of the donors was identical to the sex of the recipients. The different donors and recipients were either from our SPF unit or were obtained from a small isolator-raised group of animals with HF. The composition of the SPF flora and HF is given in Table 4.6. The composition of the aerobic part of both floras was checked twice weekly and did not change during the experiments. According to the microbiological status of the donors and the recipients, there were four different experimental groups. The (microbiological) origin of the donors and the recipients, as well as the distribution of both sexes in the different experimental groups is given in Table 4.7. The different experimental groups were: HF recipients of SPF donor bone marrow (SPF→HF), HF recipients of HF donor bone marrow (HF→HF), SPF recipients of SPF donor bone marrow (SPF→SPF), and SPF recipients of HF donor bone marrow (HF→SPF).

Table 4.6. Composition of the different microfloras

SPF flora	Houston flora
Anaerobic microflora:	
CRF flora (not defined)	Not defined
Aerobic microflora:	
Streptococcus faecalis (7173711*)	*Streptococcus faecium* (7355510*)
	Streptococcus faecium (7317550*)
Staphylococcus aureus (6726153**)	*Staphylococcus xylosus* (6736552**)
Staphylococcus epidermidis (6706133**)	*Staphylococcus haemolyticus* (6632171**)
Escherichia coli (5144572***)	*Escherichia coli* (5144572***)
	Escherichia coli (5144532***)
	Proteus mirabilis (0536000***)
	Proteus mirabilis (0534000***)
	Pasteurella pneumotropica (1220000****)[a]

* Biotype (API 20 Strep; API System, Montalieu-Vercieu, France).
** Biotype (API Staph).
*** Biotype (API 20 E).
**** Biotype (API 20 NE).
[a] Isolated from nasal washings only.

Table 4.7. Microbiological status of the donors and recipients in the different experimental groups

Experimental group	Flora donor	Flora recipient	Number of recipients	
			Males	Females
SPF → HF	SPF	HF	14	11
HF → HF	HF	HF	12	8
SPF → SPF	SPF	SPF	25	–
HF → SPF	HF	SPF	–	19

In both series of experiments, all recipients were housed under conditions of strict reverse isolation in a laminar crossflow isolator to prevent contamination of the animals with any new micro-organisms [31]. The animals received autoclaved (10 min, 134°C) AM-II food pellets (Hope Farms B.V., Woerden, The Netherlands) and acidified (pH 2.8) sterile drinking water.

The recipients were lethally (9.0 Gy) irradiated as a conditioning for BMT. The next day, they were injected intravenously with 10^7 bone marrow cells obtained from the femurs of C57BL/Rij donor mice. Irradiation of the mice and transplantation of the bone marrow cells were performed under conditions of strict reverse isolation.

No significant difference in survival (about 90%) was observed between HF→HF, SPF→SPF and HF→SPF recipients (p>0.10), but the differences in survival between the SPF→HF recipients (about 20%) and the recipients of the three other experimental groups were statistically significant ($p<0.01$).

In the second series of experiments, a mitigation of GvHD (identical to that in the HF→HF, SPF→SPF and HF→SPF recipients) was observed in the completely decontaminated HF C3H/Law recipients of SPF C57BL/Rij bone marrow, while no effect of selective GID of the HF recipients was seen.

In the first series of experiments, lethal GvHD occurred in only one of the four transplanted groups of mice. This was the group of "conventional" (HF) recipients of SPF bone marrow (SPF→HF), in which severe diarrhoea was also observed and which showed the most distinct histopathological lesions. When the donors and recipients of the bone marrow were both HF carrying animals (HF→HF), more than 90% of the recipients survived after BMT. In the two other groups in which the recipients were SPF (i.e. SPF→SPF and HF→SPF), more than 90% of the recipients survived. This is comparable with observations made when recipients of allogeneic H-2 mismatched bone marrow are germ-free or have been decontaminated [10]. The results show that mortality due to GvHD after allogeneic H-2 mismatched BMT is significantly prevented when there is identity between the microflora of the donor and recipient (HF→HF and SPF→SPF) or when the recipients do not harbour any aerobic micro-organisms other than those present in the donors (HF→SPF).

In previous publications it was hypothesized that certain bacteria may play a role in the induction of delayed type GvHD after such allogeneic BMT because they carry antigens that are cross-reactive with epithelial antigens. After transplantation of allogeneic bone marrow from a donor which was matched for the major histocompatibility complex (MHC), a mitigation of GvHD was observed in selectively decontaminated dogs [21], which confirmed the findings in mice [10] suggesting that Enterobacteriaceae might play a role in the induction and severity of GvHD. The observation that selective GID of HF recipients did not mitigate GvHD, in contrast to selective GID or HF recipients, does not support this hypothesis. Therefore, it has to be assumed that micro-organisms other than Enterobacteriaceae are involved in the induction of GvHD. It is likely that bacteria belonging to the complex anaerobic microflora are responsible for the induction of GvHD. Foo and Lee found that antigenic cross-reaction exists between mouse intestine and a *Bacteroïdes* species, one of the anaerobic members of the autochthonous microflora of the rodent gastrointestinal tract [32]. This is in accordance with the observations that selectively decontaminated human recipients of allogeneic bone marrow developed significantly more ⩾ grade II acute GvHD than completely decontaminated patients. The most striking difference between these two groups of patients was the presence of an anaerobic microflora in the first group of recipients [28].

The high mortality due to GvHD in HF recipients of SPF bone marrow in the first series of experiments can be explained by a double mechanism in evoking GvHD after allogeneic BMT. The first mechanism is the reaction of donor type T lymphocytes against histocompatibility antigens. This graft-versus-host reaction (GvHR) gives rise to minimal intestinal lesions and subsequent limited mortality, as was the case in the completely decontaminated recipients. If, however, the flora of the recipient of the allogeneic bone marrow contains anaerobic bacteria which do not belong to the indigenous flora of the donor, and which carry antigens cross-reactive with tissue antigens of the host, the donor T lymphocytes are activated by these cross-reactive bacterial antigens and become reactive against host tissues. This results in a clinically and histologically more severe GvHR, leading to the death of the majority of the recipients.

Summary

One of the major complications of allogeneic BMT is GvHD, which is caused by donor type lymphocytes which react against the recipient's tissues.

An important factor influencing GvHD is the recipient's gastrointestinal microflora. This was originally observed in gnotobiotic mice. Infusion of 10^7 H-2 incompatible bone marrow cells into lethally irradiated (9.0 Gy X-rays) conventional mice results in a late onset type GvHD which does not give rise to symptoms until 3 weeks after BMT. This disease kills the majority of the recipients during the next 2 months. Mortality attributable to this GvHD can be completely prevented if the recipients are germ-free mice, or when they are conventional animals that have been subjected to complete or selective GID by means of orally administered nonabsorbable antibiotics prior to transplantation. In studies with rhesus monkeys, a mitigating effect of complete GID of the recipient on GvHD after transplantation with incompletely T lymphocyte depleted bone marrow grafts from matched as well as mismatched donors was repeatedly demonstrated.

To explain these observations it was postulated that GvHD is the result of T cell stimulation, or of activation by cross-reactive antigens present on enteric bacteria and the recipient's epithelium, which is the main target in GvHD. This hypothesis was supported by results with transplants of fetal gut fragments in mice. Fetal gut of CBA and (CBA × C57BL) F1 hybrid mice was grafted subcutaneously in (CBA × C57BL) F1 hybrid mice. Following establishment of the grafts, the carriers were lethally irradiated and grafted with CBA bone marrow and spleen cells to induce acute GvHD. This system permits the histological grading of GvHD lesions in the normal incompatible conventional gut, as well as in bacteria-free incompatible and germ-free compatible guts in the same animals simultaneously. Results show that the presence of microflora at a distant site is capable of magnifying GvHD lesions in germ-free guts and that the effective lymphocytes act in a tissue specific way.

In experiments using mice, the mechanism responsible for the influence of the microflora on GvHD after allogeneic BMT was investigated. These studies also showed that GvHD can be induced by activated T lymphocytes from donor origin reacting against bacterial antigens which might be cross-reactive with the recipient's epithelial tissue antigens. Activation of these T lymphocytes is confined to antigens of certain bacteria of the recipient which are not present in the indigenous microflora of the donor mice. These bacteria most likely belong to the anaerobic flora of the recipient. The latter hypothesis is strongly supported by the observation in human patients that, in contrast to complete GID, selective decontamination of the gastrointestinal tract did not exert any beneficial effect on moderately severe to severe GvHD after transplantation with MHC-matched sibling donor bone marrow grafts.

References

1. Löwenberg B, Gale RP (1985) Aplastic anaemia. In: van Bekkum DW, Löwenberg B (eds) Bone marrow transplantation, biological mechanism and clinical practice. Marcel Dekker, New York, pp 409–433
2. Dicke KA, Zander AR, Vellekoop L, Spitzer G, Verma DS (1985) The treatment of leukemia. In: van Bekkum DW, Löwenberg B (eds) Bone marrow transplantation, biological mechanisms and clinical practice. Marcel Dekker, New York, pp 435–474
3. Dooren LJ, Vossen JM (1985) Severe combined immunodeficiency: Reconstitution of the immune system following bone marrow transplantation. In: van Bekkum DW, Löwenberg B (eds) Bone marrow transplantation, biological mechanisms and clinical practice. Marcel Dekker, New York, pp 351–381
4. Barranger JA (1984) Marrow transplantation in genetic disease. N Engl J Med 311:1629–1631
5. Gale RP, Bortin MM, van Bekkum DW, Biggs JC, Dicke KA, Gluckman E, Good RA, Hoffmann RG, Kay HEM, Kersey JH, Marmont A, Masaoka T, Rimm A, van Rood JJ, Zwaan FE (1987) Risk factors for acute graft-versus-host disease. Br J Haematol 67:397–406
6. Uphoff DE, Law LW (1958) Genetic factors influencing irradiation protection by bone marrow. II. The histocompatibility-2 (H-2) locus. J Natl Cancer Inst 20:617–624
7. van Bekkum DW (1964) The selective elimination of immunologically competent cells from bone marrow and lymphatic cell mixtures. I. Effect of storage at 4°C. Transplantation 2:393–404
8. van Bekkum DW (1972) Use and abuse of hemopoietic cell grafts in immune deficiency diseases. Transpl Rev 9:3–53
9. van Bekkum DW, de Vries MJ (1967) Radiation chimeras. Academic Press, New York
10. van Bekkum DW, Roodenburg J, Heidt PJ, van der Waaij D (1974) Mitigation of secondary disease of allogeneic mouse radiation chimeras by modification of the intestinal microflora. J Natl Cancer Inst 52:401–404
11. Jones JM, Wilson R, Bealmear PM (1971) Mortality and gross pathology of secondary disease in germfree mouse radiation chimeras. Radiat Res 45:577–580
12. Truitt RL (1978) Application of germfree techniques to the treatment of leukemia in AKR mice by allogeneic bone marrow transplantation. In: Waters H (ed) Immunotherapy. Garland STPM Press, New York, pp 431–452 (The handbook of cancer immunology, vol 5)
13. Heit H, Wilson R, Fliedner TM, Kohne E (1973) Mortality of secondary disease in antibiotic treated mouse radiation chimeras. In: Heneghan JB (ed) Germfree research: biological effects of gnotobiotic environment. Academic Press, New York, pp 477–483
14. van Bekkum DW (1977) Bone marrow transplantation. Transpl Proc 9:147–154
15. van Bekkum DW, Knaan S (1977) Role of bacterial flora in development of intestinal lesions from graft versus host reaction. J Natl Cancer Inst 58:787–790
16. Wagemaker G (1986) Problems and prospects of histoincompatible bone marrow transplantation studied in rhesus monkeys. In: Hagenbeek A, Löwenberg B (eds) Minimal residual disease in acute leukemia. Martinus Nijhoff, Dordrecht, pp 363–373
17. Heidt PJ, Wagemaker G, Knaan-Shanzer S, van Bekkum DW (1981) Two distinct types of late onset graft-versus-host disease after bone marrow transplantation in lethally irradiated mice. Transplantation 32:263–264
18. Heidt PJ (1989) Gnotobiotics and bone marrow transplantation: experimental and clinical studies. Thesis, University of Leiden
19. Broerse JJ, van Bekkum DW, Hollander CF, Davids JAG (1978) Mortality of monkeys after exposure to fission neutrons and the effect of autologous bone marrow transplantation. Int J Radiat Biol 34:253–264
20. Dicke KA, van Bekkum DW (1971) Allogeneic bone marrow transplantation after elimination of immunocompetent cells by means of density gradient centrifugation. Transpl Proc 3:666–668
21. Heidt PJ, Vriesendorp HM, Zurcher C, de Groot CW, Klapwijk WM (1981) Selective gastrointestinal decontamination in dogs undergoing allogeneic bone marrow transplantation. In: Sasaki S, Ozawa A, Hashimoto K (eds) Recent advances in germfree research. Tokai University Press, Tokyo, pp 579–582
22. Storb R, Prentice RL, Buckner CD, Clift RA, Appelbaum F, Deeg J, Doney K, Hansen JA, Mason M, Sanders JE, Singer J, Sullivan KM, Witherspoon RP, Thomas ED (1983) Graft-versus-host disease and survival in patients with aplastic anemia treated by marrow grafts from HLA-identical siblings. N Engl J Med 308:302–307
23. Mahmoud HK, Schaefer UW, Schüning F, Schmidt CG, Bamberg M, Haralambie E, Linzen-

meier G, Hantschke D, Grosse-Wilde H, Luboldt W, Richter HJ (1984) Laminar air flow versus barrier nursing in marrow transplant recipients. Blut 49:375–381
24. Schmeiser T, Kurrle E, Arnold R, Heit W, Krieger D, Kubanek B, Heimpel H (1984) Application of antimicrobial prophylactic treatment to the prevention of infection and graft-versus-host disease in allogeneic bone marrow transplantation (BMT). Exp Haematol 12 (suppl 15):105–106
25. Storb R, Thomas ED (1985) Graft-versus-host disease in dog and man: the Seattle experience. Immunol Rev 88:215–238
26. Leblond V, Belanger C, Dreyfus F, Brunet F, Gabarre J, Asselain B, Binet L (1987) Interest of laminar air flow for prevention of GvHD and infections in BMT for leukaemia and lymphoma. Bone Marrow Transpl 2 (suppl 1):181
27. Skinhøj P, Jacobsen N, Høiby N, Faber V and the Copenhagen Bone Marrow Transplant Group (1987) Strict protective isolation in allogeneic bone marrow transplantation; effect on infectious complications, fever and graft versus host disease. Scand J Infect Dis 19:91–96
28. Vossen JM, Heidt PJ (1988) Effect of gnotobiotic measures on infection and graft-versus-host disease in children treated with bone marrow transplantation for leukemia or severe aplastic anaemia. In: Gnotobiology and its applications. Proceedings of the IXth International Symposium on Gnotobiology. Fondation Marcel Mérieux, Lyon Cedex, pp 140–143
29. van der Waaij D, Vossen JM, Korthals-Altes C (1973) Patient isolators designed in the Netherlands. In: Heneghan JB (ed) Germfree research. Biological effect of gnotobiotic environment. Academic Press, New York, pp 31–36
30. Thomas ED, Storb R, Clift RA, Fever A, Johnson LF, Neiman PE, Lerner KG, Glucksberg H, Buckner CD (1975) Bone marrow transplantation. N Engl J Med 292:895–902
31. van der Waaij D, Andreas AH (1971) Prevention of airborne contamination and cross-contamination in germ-free mice by laminar flow. J Hyg, Camb 69:83–89
32. Foo MC, Lee A (1974) Antigenic cross-reaction between mouse intestine and a member of the autochthonous microflora. Infect Immunol 9:1066–1069

Section II

MALIGNANCIES

Chapter 5

Human T Lymphotropic Retroviruses: Pathological Consequences of Infection

R. C. Gallo

Introduction

This volume focuses on microbes that cause disease but not by an infectious process, meaning that the effects of the microbe are very delayed, indirect, or both. Strictly speaking, the microbe is infectious but when the disease presents itself the microbe is not present or "inactive". I will describe microbes which fit the general theme but not precisely, because during disease the microbe may still be transmitted (hence "infectious") albeit often with difficulty and not in the acute fashion we usually think of an infectious process.

For many years now my colleagues and I have been interested in viruses with unusual properties, viruses which are new in man or newly discovered but which have become increasingly important in the latter part of the twentieth century. These viruses are generally difficult to transmit, because they replicate poorly or have a very restricted range of target cell types or replicate only in a very limited part of a cell's life span or all of these properties. Because of this they are often difficult to discover and isolate. Also, partly because of these restrictions they *usually* take a long time to cause disease, years for most and even decades for some, but the process may be accelerated by larger doses of virus at the time of infection or by co-factors. When they do produce disease it is usually serious disease–neoplasias, immune disorders, and progressive neurological abnormalities. Although difficult to transmit and geographically well localized they have more recently become global. Two factors converged to account for this. The first is the nature of these viruses which lead to life-long infection and infection transmission for years. The second is human social change, particularly the post-World War II expansion of travel, new contacts as well as the widespread international use of blood and blood products combined with the destructive habit of intravenous drug abuse. What might have remained rare and remote is becoming relatively common and global. I include in these viruses some of the hepatitis viruses, especially hepatitis B virus (HBV); some papilloma viruses; some herpes viruses (with the exception that they are more readily transmitted); and most importantly and most fittingly all of the human retroviruses. It is also likely that other categories of viruses with properties similar to these are yet to be discovered.

Human Retroviruses: Some General Comments

It is of interest that all known human viruses that chiefly target the central cell of the immune system, the $CD4^+$ T cell, were discovered within the last decade (Fig. 5.1). There are five: human T cell leukaemia viruses I and II (HTLV-I, HTLV-II), discovered in 1979–1980 and 1981–1982 respectively; human immune deficiency virus 1 and 2 (HIV-1, HIV-2), discovered in 1983–1984 and 1985–1986, respectively; and human herpesvirus-6 (HHV-6), discovered in 1986. The first four are, of course, retroviruses; the fifth, discovered by Salahuddin and Ablashi, is a new human herpesvirus (previously called human B cell leukaemia virus, HBLV) [1–3]. The discovery of the retroviruses came in this decade chiefly because of our capacity to grow target T cells with interleukin-2, (IL-2) and because of the development of sensitive assays for their detection. Although these viruses may infect one or more other cell types, the $CD4^+$ T helper cell is the primary target for each. The receptor on the cell surface differs. HIV-1 and HIV-2, of course, use the CD4 molecule itself while the receptor for HTLV-I, HTLV-II and HHV-6 is unknown but is not CD4. HTLV-I and HTLV-II immunologically cross-react and share about 50% nucleotide sequence similarity. HIV-1 and HIV-2 immunologically cross-react and also share about 50% nucleotide similarity. However, the HTLVs and HIVs do not cross-react with each other and have little similarity in their nucleotide sequences. In fact, by morphology, genome organization, and targeting also of non-T cells, i.e. macrophage–monocytes and their lineage relatives such as Langerhans cells, follicular dendritic cells of the lymph nodes, and microglial cells of brain, the HIVs are categorized in the subfamily of retroviruses known as the lentiretroviruses while the HTLVs belong to the oncorna or type-C retrovirus family. Oddly, however, the HTLVs and HIVs do share important biological properties with each other that are often not common with their respective family relatives in animals. In addition to the T4 tropism, a common manner of transmission (blood, sex, and mother-to-infant in milk or by a perinatal or congenital infection), each has close relatives among African primates and a few special features of virus replication.

HTLV-I and T Cell Leukaemias and Lymphomas

HTLV-I is the cause of certain forms of adult T cell leukaemias and lymphomas [4,5]. The linkage is not difficult to establish because both the virus and the

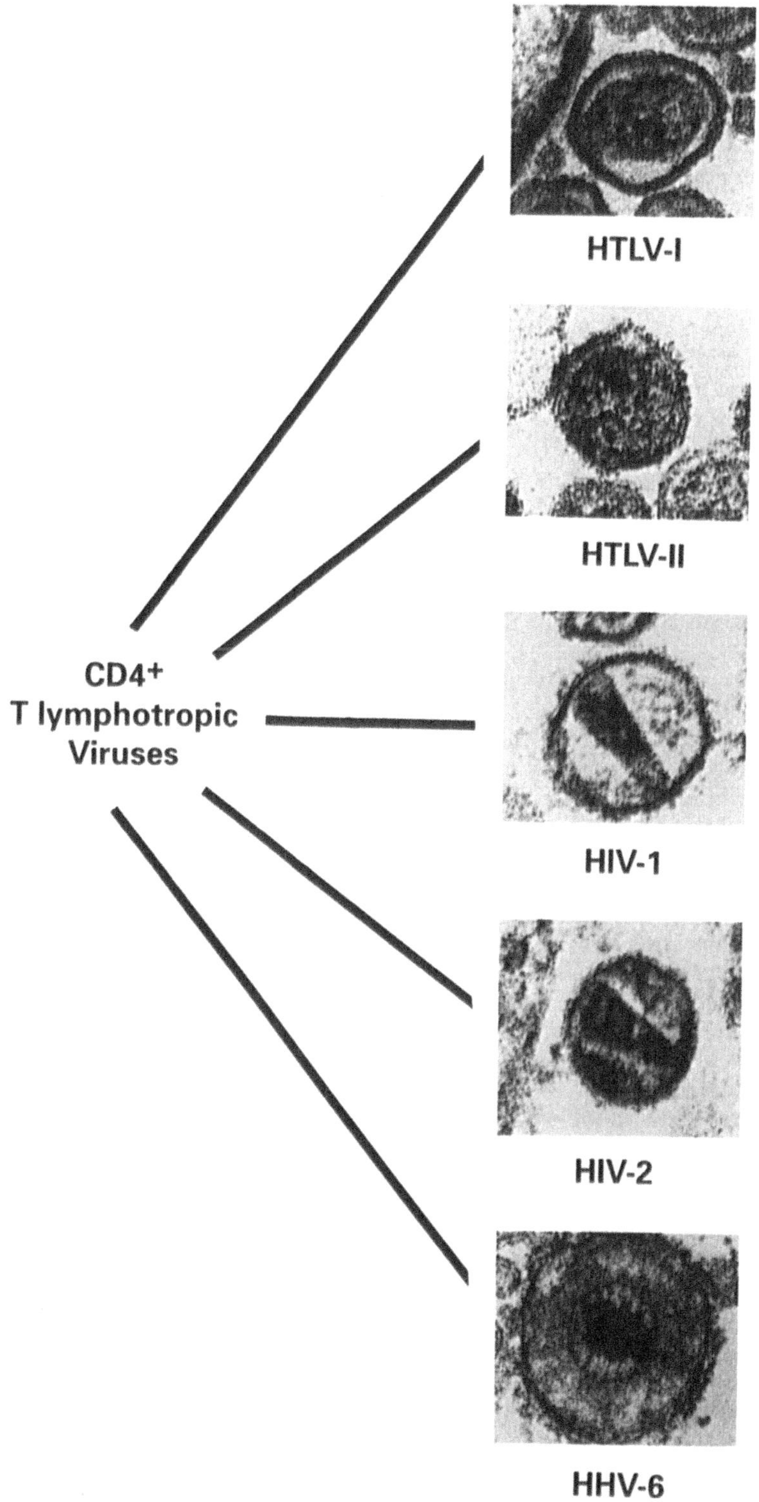

Fig. 5.1. Five human viruses which are primarily tropic for CD4$^+$ T lymphocytes. The mature cell free virions are shown.

disease are uncommon, making the epidemiological connection rather straightforward. Moreover, in these neoplasias the DNA provirus is present and integrated in a clonal fashion on the tumour, proving that the virus was present in the first transformed cell and did not come later as a "passenger" virus. The classical HTLV-I malignancy is a very aggressive tumour with death usually occurring within 6 months of diagnosis. Often (50%) there are skin changes such as tumour cell infiltrates and a high blood calcium (due to the release of lymphokines from infected cells which activate osteoclasts, leading to lytic bone lesions and hypercalcaemia). Opportunistic infections are common. In a typical case a leukaemia is common and the T cells have a characteristic appearance with hyperlobulated nuclei. The first patient from which a retrovirus was ever isolated was exactly this type. Like most, he was relatively young and had no medical or social history of note, living all of his life in a small town in Alabama. Despite the fact that disease progression is rapid, the time from virus infection to disease is usually very long (years to decades). The malignant cell is almost always of the infected T4 lymphocyte.

Similar to but clinically and pathologically distinguishable from the "classic" adult T cell leukaemia/lymphoma (ATL) are other T cell malignancies of adults which are caused by HTLV-I. These are a small percentage of T cell chronic lymphocytic leukaemia (T-CLL), T-cell non-Hodgkin's lymphoma, and mycosis fungoides. The common feature in these and the ATL cases are the T4 cell type and the abundant expression of IL-2 receptors. In regions where HTLV-I is prevalent, e.g. the Caribbean, the percent of HTLV-I linkage with these other clinicopathological diagnoses is much higher. Like Epstein–Barr virus (EBV), the HTLVs immortalize their target T cells in vitro, and presumably this is a major effect of the virus in bringing out the malignancy. However, unlike EBV, the HTLV induced malignancy is not geographically facilitated. Wherever HTLV-I is present it produces these kinds of tumours. Unlike EBV lymphomas (Burkitt's) no additional environmental factors must be postulated, and unlike EBV lymphomas none have been found.

The incidence of HTLV-I induced leukaemia/lymphoma is about one in a few hundred infected people per lifetime. This is similar to the animal leukaemia viruses such as feline leukaemia virus. In such animals it has been shown that given the right age, dose of virus, and route of infection, the incidence of disease can approximate 100%.

We know much less about HTLV-II and linkage to disease. When found, it has been in more chronic forms of adult T4 cell laukaemias.

The Mechanisms of HTLV-I Replication and Induction of Leukaemia

Studies with the HTLVs were not only of interest because of their role in human disease or for establishing the existence of human retroviruses. Out of basic studies of these viruses also came novel genes, mechanisms for replication, and mechanisms for inducing leukaemia. Animal retroviruses were known to have only three genes: *gag* for the core proteins, *pol* for the viral enzymes, and *env* for the viral envelopes. Molecular analyses of the HTLV-I and HTLV-II genomes revealed two extra genes at the 3′ end of the genome, now called *tax* and *rex* (Fig. 5.2). These genes code for nuclear regulatory proteins. Without them HTLVs can

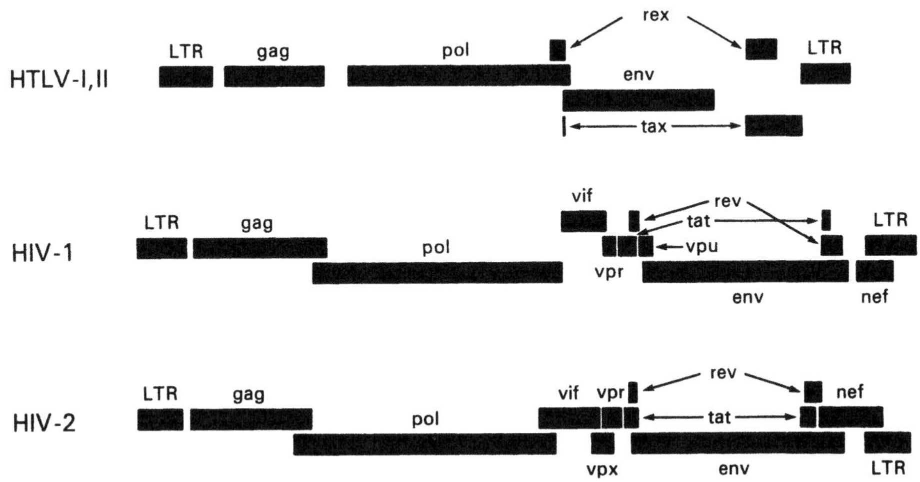

Fig. 5.2. Genomic structure of human retroviruses. Viral genes are shown as open reading frames. In addition to the usual retroviral genes (*gag*, core protein; *pol*, polymerase or reverse transcriptase; and *env*, envelope), each virus encodes one or more gene products that are important in regulating viral replication. Some of these mRNAs are formed by a double spliced mechanism shown by the *arrows*. The *LTR* (long terminal repeat) regions contain viral sequences important in regulation and replication.

not replicate. *Tax* and *rex* are early genes, another novelty for retroviruses. Only when they are expressed are the viral structural proteins made. Moreover, *tax* and *rex* have introduced some novel molecular biological functional features. *Tax* appears to regulate the formation of viral RNA transcribed from the integrated DNA provirus. The molecular function of *rex* is to augment the maturation and transport of mRNA for *gag*, *pol* and *env* – the structural proteins – while having an opposite effect on its own mRNA and that of *tat*. As far as we know, the earlier events in the replication life cycle are not unlike the known steps with animal retroviruses: binding of the viral envelope to a cell surface receptor, penetration of the cell by the core of the virus, reverse transcription with eventual formation of the double-stranded linear DNA form which enters the nucleus and integrates into the host chromosomal DNA. It is at this stage (Fig. 5.3) that the steps begin to differ. Infection of a permissive cell by previously studied animal retroviruses usually leads to a continual transition from the incoming phase to the outgoing phase. However, with the human retroviruses after integration the provirus is usually "silent".

We do not know what activates the expression of *tax*, but only after this happens will the rest of the viral genes be expressed. *Tax* itself may activate T cell genes, probably through NFκB related sequences. Once the viral RNAs are made and transported subsequent events follow those of any retrovirus (protein synthesis and cleavage maturation, assembly of viral proteins with genomic RNA, budding and release of the virus from the cell membrane).

Other studies indicate that *tax* is critical to the immortalization of T cells by the HTLVs and for induction of leukaemia. Included in the cellular genes activated by *tax* are IL-2 and the interleukin-2 receptor (IL-2R). We think this leads to

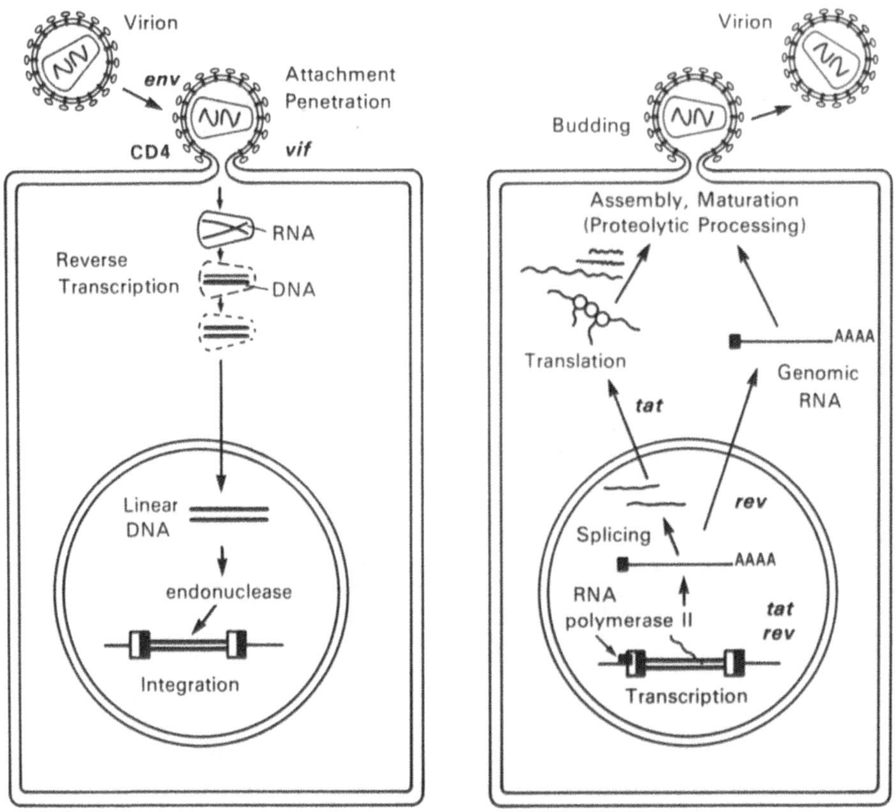

Fig. 5.3. Life cycle of HIV. Virus enters the cell by the interaction of envelope (*env*) protein and the CD4 molecule. Once uncoated in the cytoplasm, a double-stranded DNA copy of the viral single-stranded RNA is made by viral reverse transcriptase. The DNA form of the virus then migrates to the cell nucleus where it integrates into the host cell DNA. In the form of the provirus it may remain "silent" or express viral genes leading to the assembly of new virus. The life cycle is completed by the budding from the cell surface.

autocrine- and paracrine-stimulated T cell proliferation. This activation by *tax* is called *trans*-activation. It differs, then, from the conventional way animal retroviruses are thought to induce leukaemia, i.e. by a *cis* activation of a proto-oncogene. This means HTLV need not integrate near a proto-oncogene to cause leukaemia and may explain why leukaemia can occur with so little virus compared to most animal systems. A polyclonal T cell proliferation can often be shown to precede the leukaemia which is probably the in vivo analogue of paracrine- and autocrine-stimulated growth. Cells expressing viral proteins, however, are probably cleared by immune response and we are back to the stage of cells containing "silent" proviruses. I think this is a recurrent cycle occurring for years, and only when a second, perhaps chance, genetic event occurs will the clonal tumour develop. Clinically this phase may be seen and is called "smouldering" ATL. A third genetic change may be necessary for the acute aspect of the disease. However, no specific chromosomal changes or alterations in proto-oncogenes have been found.

HTLV-I and Neurological Disease

A few years ago deThé and co-workers discovered a link between HTLV-I and a serious neurological disease which is distantly similar to multiple sclerosis. This was rapidly confirmed and extended in several parts of the world. This disease has been called Jamaican neuropathy (in Jamaica), tropical spastic paraparesis (in Africa) and variants of multiple sclerosis in most other places, and in Japan it is called HTLV-I associated myeloencephalopathy or HAM. The mechanisms for this myeloencephalopathy phenomenon are indeed obscure, particularly in view of the fact that no one has ever shown that HTLV-I infects cells of the central nervous system (CNS). Moreover, since several patients have been found with both ATL and HAM, it does not appear to be due to a different strain of the virus. One probably very significant clue is that people with this disease have much higher antibody titres and appear to have less control of viral replication. An autoimmune mechanism is currently an attractive idea.

HIV and the Pathogenesis of AIDS

The major manifestations of AIDS consist of:

1. Immune impairment and subsequent opportunistic infections;
2. Infection of the brain and its sequelae;
3. Abnormal growth of some cells, notably the development of Kaposi's sarcoma (KS) and B cell lymphoma;
4. Miscellaneous diseases due to pathogens such as tuberculosis (TB), the HTLVs, papilloma viruses, HBV, and herpes viruses. Sometimes the lifestyle of HIV infected people leads to a greater risk of exposure to these microbes as well as those which are opportunistic while at increased risk as a result of immune impairment.

Only the first point and KS will be discussed here.

Immune Impairment

The essence of HIV induced immune abnormalities involves five events:

1. The targeting of $CD4^+$ cells, by a remarkable affinity between a component of the gp120 protein of the HIV envelope and the outer portion of CD4. This phenomenon targets HIV to the heart of the immune system – the T4 helper lymphocytes and the monocyte-macrophages and related cells.

2. Special features of its genome and mode of replication. Like the HTLVs, the genome of the HIVs has extra regulatory genes, but instead of two, there are at least eight. Despite this much greater complexity it is only two of these, called *tat* and *rev*, that are essential to virus replication, and these are functionally similar to *tax* and *rex* of the HTLVs. Like the HTLVs, infection of T4 cells by the HIVs initially does not usually go past integration. The silent provirus in this case

requires immune stimulation and cytokines to be activated. When the T cell activation genes are expressed the HIV provirus is expressed and the virus replicates.

3. The cytopathic effect. It is very unusual for retroviruses to kill their targets, but an HIV infected T4 cell activated to release virus soon dies. Also, cells which do not form virus but express gp120 on their surface can bind to CD4 of uninfected adjacent cells leading to giant multinucleated cell (syncytia) formation. Such cells do not function and die prematurely. These direct effects must have a major role in the T4 depletion in AIDS, but indirect mechanisms may also be operative. For instance, gp120 binding to CD4 of uninfected cells attracts antibodies to gp120 which in turn brings in killer cells to the antibody Fc component. In this way uninfected cells can be killed, as shown by Weinhold et al. [6]. Still other mechanisms may be involved such as the incorporation and processing of gp120 by activated uninfected T4 cells which then present it with HLA and become cytotoxic lymphocyte (CTL) targets [7]. There are also suggestions of molecular mimicry (an envelope component with HLA-DR) and of viral proteins inhibiting T cell function (e.g. gp120 inhibition of T cell activation). Finally, the possibility of other "enhancing" viruses should probably be considered. In this regard HHV-6 can also kill T4 cells, but it is normally kept under control by the immune system. If replication of HHV-6 increases it might be responsible for more T4 cell killing. In addition, HHV-6 can activate HIV-1 [8].

4. A fourth essential aspect of this problem is the occurrence of infectious intracellular HIV in infected macrophage. Its significance lies in the fact that when these cells are killed more infectious virus is released. In theory this should not be a problem if there is high titre of effective neutralizing antibody, but this is not always the case. This brings me to the next probable essential element of AIDS.

5. Heterogeneity of the viral genome and recurrence of variants resisting neutralization antibodies. At any one time some population of variants in an HIV infected person dominates, which may induce neutralizing antibodies, allowing other variants to emerge.

Kaposi's Sarcoma

KS occurs in association with HIV about 1000 times more frequently than in non-HIV infected people. The tumour is peculiar for several reasons. It has a multifocal and heterogenous appearance, may not be of clonal origin, has no known chromosomal abnormalities, consists of mixed cell types rather than a uniform tumour cell population, and sometimes can spontaneously regress. The lesion consists of proliferating blood vessels, fibroblasts, infiltrating white blood cells (WBC), and a cell type believed to be the tumour cell of KS, known as the spindle cell. Because of a higher incidence of KS in HIV infected homosexual men we evaluated KS tissues for the presence of other viruses, which, with HIV, might have been the inducing agent of KS. We found none, but in these studies we learned how to grow the KS spindle cells for long periods [9]. This allowed us to characterize these cells, yielding interesting results. First, studies in collaboration with J. Folkman suggest that the spindle cells are probably derived from smooth muscle. Realizing that the hallmark of KS is neoangiogenesis, we tested the cultured spindle cells for such activity. We found that the KS spindle cells release large amounts of neoangiogenic factors. They synthesize large amounts of fibroblast growth factor, (bFGF) and IL-1 [10], which may account for part or all of this activity. The spindle cells also release granulyte-macrophage colony stimulating factor (GM-CSF), platelet derived growth factor (PDGF), transforming growth factor (TGF), and growth factors for themselves. The most interesting result, however, was the induction of KS-like lesions of mouse origin in nude mice 6–10 days after inoculation of the human spindle cells into these animals. These KS-like lesions regress after the spindle cells die out in the animal.

In recent months we turned our attention to the source of growth factor which allowed us to grow the spindle cells in the first place. One major source of growth factor was the conditioned medium of HIV infected T4 cells. Through a series of experiments we discovered that the active factor is *tat*. This is a remarkable result. *Tat*, a viral *trans*-acting regulator of viral RNA formation, is released into the medium of early infected T4 cells. *Tat* does not have a signal peptide and the release is not related to cell degeneration. We have shown that the active factor is *tat* by removing its affinity with purified anti-*tat* antibodies, and directly by using recombinant *tat* proteins.

Human retroviruses can stimulate growth by two very different mechanisms: a direct transforming effect by HTLV of a target cell and an indirect effect as in the case of HIV. Here the target T4 cell or macrophage does not become the tumour but releases a novel growth factor (*tat*) which stimulates other cells (the spindle cell) to grow and to produce cytokines leading to the KS lesion.

Acknowledgements. The author wishes to thank Dr Howard Streicher for critical reading of the manuscript, Dr Lata Nerurkar for helping with the figures, and Ms Anna Mazzuca for editorial assistance.

References

1. Salahuddin SZ, Ablashi DV, Markham PD, Josephs SF, Sturzenegger S, Kaplan M, Halligan G, Biberfeld P, Wong-Staal F, Kramarsky B, Gallo RC (1986) Isolation of a new virus, HBLV, in patients with lymphoproliferative disorders. Science 234:596·601
2. Gallo RC (1987) The AIDS virus. Sci Am 256:47–56
3. Gallo RE, Montagnier L (1988) AIDS in 1988. Sci Am 259:41–48
4. Gallo RC (1986) The first human retrovirus. Sci Am 255:88–98
5. Blattner WA (1989) Retroviruses. In: Evans AS (ed) Viral Infections of Humans, 3rd edn. Plenum, New York, pp 545–592
6. Weinhold KJ, Matthews TJ, Ahearne PM, Langlois AJ, Lyerly HK, Tyles DS, Stine KC, Durack DT, Bolognesi DP (1988) Cellular anti-gp120 cytolytic reactivities in HIV-1 seropositive individuals. Lancet i:902–904
7. Siliciano RF, Lawton T, Knall C, Karr RW, Berman P, Gregory T, Reinherz EL (1988) Analysis of host–virus interactions in AIDS with human T cell clones specific for HIV gp120: effect of HIV genomic heterogeneity and a mechanism for $CD4^+$ cell depletion. Cell 54:561–575
8. Lusso P, Ensoli B, Markham PD, Ablashi DV, Salahuddin SZ, Tschachler E, Wong-Staal F, Gallo RC (1989) Productive dual infection of human $CD4^+$ T lymphocytes by HIV-1 and HHV-6. Nature 337:370–373

9. Nakamura S, Salahuddin SZ, Biberfeld P, Ensoli B, Markham PD, Wong-Staal F, Gallo RC (1988) Kaposi's sarcoma cells: Long-term culture with growth factor from retrovirus-infected $CD4^+$ T cells. Science 242:426–430
10. Ensoli B, Nakamura S, Salahuddin SZ, Biberfeld P, Larsson L, Beaver B, Wong-Staal F, Gallo RC (1989) AIDS–Kaposi's sarcoma-derived cells express cytokines with autocrine and paracrine growth effects. Science 243:223–226

Chapter 6

Viruses, Oncogenes and Cancer

A. J. van der Eb

Cellular Oncogenes and Cancer

The concept that cancer arises as a result of mutations in cellular genes was confirmed around 1981 by Weinberg and others. They found that DNA isolated from a human bladder carcinoma cell line could transform NIH/3T3 cells, an established cell line of mouse origin, into tumourigenic cells. When a similar experiment was carried out with DNA from normal human tissue, no such transformation was observed. This result indicated that the DNA from the tumour cells contained a gene that had acquired transforming activity, presumably due to a mutation.

The human gene responsible for the transforming activity was subsequently isolated and shown to be virtually identical to the so-called H-*ras* gene. This gene had been identified before as a retroviral oncogene that had been taken up by the virus from an animal cell. The transforming activity of the *ras* gene from the bladder carcinoma cells turned out to be caused by a single point mutation which resulted in the substitution of a single amino acid in the *ras* protein: the twelfth amino acid (glycine) had been replaced by another (valine). This was the first definitive proof that a somatic mutation can convert a normal cellular gene into a cancer gene. The mutated or "activated" *ras* gene was called "oncogene", whereas the normal unmutated *ras* gene was called "proto-oncogene". After the first isolation of this *ras* oncogene, similar mutated *ras* genes were isolated from many other tumours. In all cases the transforming activity was caused by a single point mutation, which could be not only at position 12 of the *ras* protein but also at position 61 or occasionally 13.

In addition to the *ras* gene a large number of other oncogenes have been identified. At present more than 50 oncogenes are known and the list is still growing. Some of these oncogenes are activated in a similar manner to the *ras* oncogenes. The majority, however, are activated in different ways, e.g. by the fact that they had been incorporated into the genome of retroviruses or by the circumstance that they were transcriptionally activated by the adjacent insertion of the genome of a retrovirus.

The important question that subsequently arose was how these oncogenes cause cancer. To answer that question, it was necessary to know first what the functions of the proto-oncogenes are. Data from a large number of laboratories indicate that in the normal cell many proto-oncogenes have a function in the regulation of cell proliferation or differentiation. Cell proliferation is regulated by protein molecules that usually reach cells from outside. One class of these regulatory proteins are the so-called growth factors, molecules that exert a stimulatory influence on cell proliferation. When a growth factor binds to its receptor on the cell membrane, the receptor becomes activated and starts to transmit chemical signals to the nucleus by means of a complicated signal transduction pathway. When the signals arrive in the nucleus they will stimulate the expression of a number of specific genes (e.g. *fos*, *myc*, *jun*). This stimulation is followed several hours later by DNA replication and cell division. Now, it turns out that several of the steps in the mitogenic response to growth factors are controlled by proto-oncogenes. For example, growth factors may themselves act as oncogenes and so may their receptors. Some proto-oncogenes fulfil functions in the intracellular signal transduction pathways, whereas others have been shown to belong to the genes that are activated in response to growth-factor stimulation. Furthermore, there is evidence that certain proto-oncogenes have functions in cell differentiation. Mutations in these genes, or in the DNA elements controlling their expression, may then cause deregulation of cell proliferation or prevent differentiation creating the phenotypes characteristic for cancer cells.

Thus, the discoveries of oncogenes have proven to be very powerful in generating preliminary explanations of carcinogenesis at the molecular level. A further conceptual understanding has to take into account two other facts:

1. One oncogene is generally not sufficient to convert a normal diploid cell into a cancer cell, but two or more genes appear to be involved. This is not surprising as it was already known that cancer is a multistep process.

2. There is evidence that not only dominant oncogenes play a role in carcinogenesis, such as *ras* and the other oncogenes discussed so far, but also genes that must be *in*activated before cancer can develop. These genes are called tumour-suppressor genes. The best studied example of a tumour suppressor gene is the retinoblastoma-predisposition gene which plays a role in the development of both hereditary and sporadic retinoblastomas. Retinoblastoma tumours are characterized by the fact that both alleles of the so-called retinoblastoma gene are inactivated due to deletion or mutation. The presence of one intact allele of the gene is sufficient to prevent tumour formation, hence the name tumour-suppressor gene.

The concept that one oncogene is not sufficient to convert normal cells into fully malignant cells was suggested by the following observation. As mentioned earlier, introduction of a mutated *ras* oncogene into mouse NIH/3T3 cells converts them into oncogenic cells in a single step. In contrast, introduction of a *ras* oncogene into primary or secondary mouse embryo cells does not lead to oncogenic transformation. When, in addition to *ras*, a second oncogene is introduced into the embryo cells, e.g. an activated *myc* oncogene or the adenovirus E1A gene, the cells become fully tumourigenic. NIH/3T3 cells differ from the mouse embryo cells in that the former are already "immortalized" and can grow indefinitely in culture, although they are unable to form tumours. In contrast, the mouse embryo cells have a limited growth potential in vitro. This suggests that embryo cells require, in addition to *ras*, a second oncogene conferring immortality, a change that had occurred spontaneously in the NIH/3T3 cells. Although the concept of collaborating oncogenes is in agreement with the well-known multistep nature of carcinogenesis, and *ras* is the most common oncogene in human tumours, only a few examples of tumours are known that harbour two collaborating oncogenes, such as *ras* and *myc*. In fact, not many tumours are known harbouring *myc* genes that are activated by mutations. In contrast to the situation with *ras*, activation of *myc* does not involve mutations in the *myc* protein which result in altered biochemical properties, but rather is due to DNA rearrangements causing enhanced or deregulated expression of the *myc* gene. However, deregulated expression of *myc* may also be achieved by genetic changes in genes other than the *myc* gene itself. In normal cells expression of *myc* is tightly coupled with growth. In quiescent fibroblasts *myc* is expressed at low levels, but when the cells are induced to grow (by growth factors) its expression is substantially increased. It is conceivable, however, that *myc* expression is not only dependent on positive regulation by growth factors but also on negative regulation by antimitogenic stimuli. If this assumption is correct, then constitutive *myc* expression can also be brought about by mutation or inactivation of the gene(s) that normally suppress *myc* activity. In this way *myc* can become "activated" without being altered itself. Since according to this model *myc* activation would be dependent on inactivation or loss of both alleles of a gene, it is tempting to speculate that such *myc*-regulating genes may actually belong to the category of tumour-suppressor genes. References about oncogenes can be found in [1–4].

The Role of *Ras* Oncogenes in Human Neoplasia

While little is known about the contribution of *myc*-like oncogenes in human cancer, more information is available about the role of mutant *ras* oncogenes. Studies by J. L. Bos in our laboratory have shown that the three known *ras* oncogenes occur in a significant proportion of human tumours. Table 6.1 shows that mutant *ras* genes are found in a variety of cancers, in frequencies ranging from 6%–90%, but that they are completely absent in others. From these data a number of conclusions can be drawn:

1. Within a particular category, tumours can be found with and without mutant *ras* genes. The phenotypical properties of tumours with and without *ras* oncogenes are the same. This can be explained in at least two ways: (a) the tumours without *ras* oncogenes contain other as yet unknown oncogenes with the same effect as *ras*; (b) mutant *ras* genes have no effect on tumour phenotype. The latter explanation is not very likely since mutant *ras* genes have strong transforming potential when introduced into NIH/3T3 cells.

Table 6.1. Incidence of *ras* mutations in human tumours. The number of individual tumours tested in each category is more than 20, usually more than 30, except for large cell lung carcinoma (10) and seminoma (14) [5]

Tumour	Tumours with *ras* mutations (%)	Preponderant *ras* gene
Breast adenocarcinoma	0	–
Ovary carcinoma	0	–
Cervical	0	–
Oesophagal	0	–
Glioblastoma	0	–
Neuroblastoma	0	–
Stomach carcinoma	0	–
Lung:		
Squamous cell carcinoma	0	–
Large cell carcinoma	0	–
Adenocarcinoma	30	K-*ras*
Colon:		
Adenoma	50	K-*ras*
Adenocarcinoma	50	K-*ras*
Pancreas carcinoma	90	K-*ras*
Seminoma	40	K-, N-*ras*
Melanoma	20	N-*ras*
Bladder carcinoma	6	H-*ras*
Myeloid disorders:		
Myelodysplastic syndrome	30	N-*ras*
Acute myeloid leukaemia	30	N-*ras*
Chronic myeloid leukaemia	10	–
Lymphoid disorders:		
Acute lymphoid leukaemia	0	–
Non-Hodgkin's lymphomas	0	–
Hodgkin's lymphomas	0	–
Paediatric leukaemias:		
Myeloid	30	N-*ras*
Lymphoid	10	N-*ras*

2. Another striking observation is that there exists a distinct specificity in the activation of particular *ras* genes in particular tumours. Human cells contain three related *ras* genes, K-*ras*, N-*ras* and H-*ras*. As can be seen in Table 6.1, colon tumours contain predominantly K-*ras* oncogenes, as do the majority of the other tumours. In contrast, N-*ras* is the most abundant *ras* oncogene in leukaemias. As yet, we do not know the cause of this specificity.

3. An interesting conclusion can be drawn from the results obtained with the colon tumours (collaboration with B. Vogelstein, Johns Hopkins University). It appeared that both the non-malignant adenomas and the malignant carcinomas have *ras* mutations in 50% of the cases. As the adenomas are considered to be the precursors of the carcinomas, these data indicate that *ras* activation is an early event and that progression to carcinoma needs additional step(s). However, occasionally the mutation of *ras* can also be a late event, suggesting that there may be no specific order in the activation of *ras* oncogenes.

Similar conclusions were drawn from studies on myelodysplastic syndrome (MDS) (collaboration with J. Yunis, Minneapolis). MDS is a myeloproliferative disorder which progresses into acute myeloid leukaemia (AML) in 30% of cases. It has been possible to follow the appearance of *ras* mutations during the various stages of the disease in individual patients. In some of the patients *ras* mutations occurred late, at the onset of the blast crisis. In other patients mutant *ras* genes were present in the early pre-leukaemic stages and during the remission after chemotherapy, even in the apparently normal lymphocytes. This indicates that the initial *ras* mutation had occurred in a pluripotent stem cell which gradually replaced most of the bone marrow. (For a summary of the role of *ras* oncogenes in human cancer, see Bos [5]).

Transformation and Oncogenesis by Human Adenoviruses

In the last part of this chapter I would like to discuss some of our work on human adenoviruses and on their use as a model to study carcinogenesis.

Adenoviruses (Ad) can be divided into two groups depending on whether they can induce tumours in hamsters. An example of an oncogenic virus is Ad12, and an example of a non-oncogenic virus is Ad5. Both oncogenic Ad12 and non-oncogenic Ad5 can morphologically transform cultured cells. When the oncogenicity of the transformed cells is tested, it turns out that cells transformed by Ad12 can form tumours in immunocompetent syngeneic animals, whereas cells transformed by Ad5 cannot. Both types of transformed cells, however, can induce tumours in nude mice (which are deficient in T cells), indicating that Ad12-transformed cells are tumourigenic in normal animals because they can evade the T cell immune defence. The viral genes responsible for oncogenic transformation were identified 15 years ago in our laboratory. They are located in a segment which is called region E1 and which consists of two subregions, E1A and E1B (Fig. 6.1). Each of these subregions codes for two proteins. Both E1A and E1B

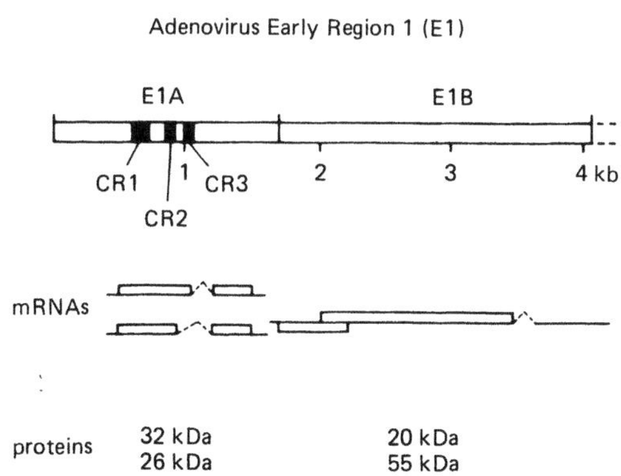

Fig. 6.1. Transforming DNA segment of human adenoviruses, called early region 1 (region E1). Region E1 consists of two subregions, *E1A* and *E1B*, which each code for two proteins. Also indicated are three conserved regions (*CR1*, *CR2* and *CR3*), which are conserved among all human adenoviruses studied. CR1 and CR2 are directly involved in cell transformation, CR3 is not.

are required for complete oncogenic transformation, in agreement with the model that spontaneous cancer also requires more than one oncogene.

In an attempt to elucidate the mechanism of adenovirus-induced transformation two main questions are being studied:

1. How do the adenovirus E1 genes convert normal cells into transformed cells with altered growth properties?
2. Why are cells transformed by Ad12 oncogenic in immunocompetent animals and cells transformed by Ad5 non-oncogenic?

In order to answer the first question we attempted to establish in which properties Ad-transformed cells differ from normal cells. Studies by Timmers et al. showed that the most obvious difference concerns the dependence on growth factors [6]. Ad-transformed cells are (almost) independent of growth factors for their in vitro proliferation, a property which they share with most other tumour cells, whereas normal cells can grow only in culture when growth factors are available. Thus, it was conceivable that adenoviruses cause transformation by interfering with the growth factor regulatory mechanism. Our current results indicate that adenovirus E1A can indeed severely alter the expression of genes that in normal cells are activated by growth factors [7,8]. The data suggest a mechanism in which E1A proteins, probably via modification of cellular proteins, drive cell proliferation autonomously, and thereby bypass the normal signal-transduction mechanism activated by growth factors. The precise molecular events are still unknown but our results indicate that one of the effects brought about by Ad E1A is an alteration of the activity of the transcription factor AP1 (collaboration with P. Herrlich, Karlsruhe; Offringa et al. [9]). AP1 represents a family of transcription factors which consist of products of the proto-oncogenes *jun* and *fos*. They play a crucial role, among others, in the control of cell proliferation. Therefore, alterations in the properties of AP1 could well contribute to deregulation of cell division.

Our final question concerns the tumourigenicity in immunocompetent animals: why are Ad12-transformed cells oncogenic in these animals and Ad5-transformed cells non-oncogenic. A possible answer to this question was provided by the discovery of Schrier and Bernards, who showed that Ad12 E1A, as opposed to Ad5 E1A, can suppress the expression of the class I major histocompatibility complex (MHC) genes in transformed cells [10,11]. Class I MHC genes occur on the cell membranes of almost all cells of the body. They play a crucial role in the recognition by cytotoxic T lymphocytes (CTL) of cells that express foreign antigens, such as viral proteins. The CTLs can only recognize and destroy such cells when the foreign antigens are presented on the cell membrane together with the class I antigens. Ad5-transformed cells express normal levels of class I antigens and the CTLs can therefore recognize and destroy them. Ad12-transformed cells do not express class I and will evade the T cell defence, so they can form a tumour. Although this explanation seemed plausible there was no formal proof that expression of class I antigens is the cause of the non-oncogenic character of Ad5-transformed cells in immunocompetent animals.

To verify this prediction and investigate whether the Ad transformation system may be suitable as a model to develop a tumour-regression protocol, Drs Melief and Kast (Amsterdam), in collaboration with our laboratory, have isolated CTL clones directed against Ad5-transformed mouse cells. The CTL clones were isolated from a mouse immunized with Ad5-transformed cells. Cloned Ad5-

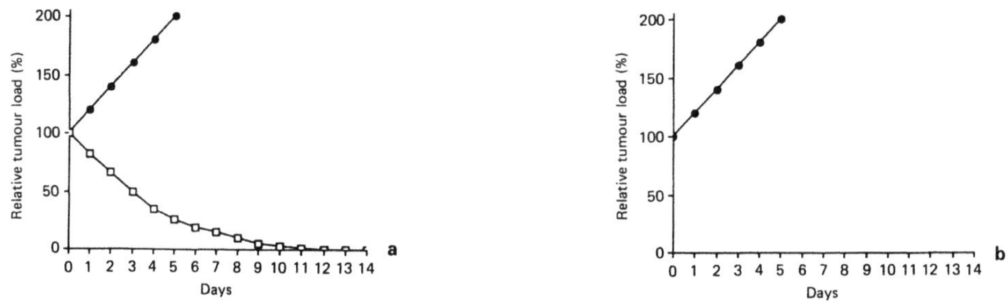

Fig. 6.2. Regression of adenovirus 5 tumours in nude mice by cytotoxic T lymphocytes (CTL) specific for adeno 5-transformed cells.
a C57BL/6 (B6) mice carrying adeno 5 B6 tumours received the following treatment: —●— no treatment, or injection with recombinant interleukin-2 (rIL-2), or with a Sendai virus-specific B6 CTL clone + rIl-2. —□— injection with an adenovirus 5-specific B6 CTL clone + rIL-2.
b B6 mice carrying adeno 12 B6 tumours were injected with an adeno 5-specific B6 CTL clone + rIL-2.

specific CTL were injected intravenously into tumour-bearing nude mice, along with 10^5 units interleukin-2 (IL-2). The tumour cells were of the same class I MHC haplotype as the CTL. The results are shown in Fig. 6.2. Treatment with IL-2 alone or a Sendai virus-specific CTL clone had no effect but the Ad5-specific CTL caused complete regression of tumours with masses of up to 10 cm^3. Ad12 tumours did not go in regression, indicating that the CTL is specific for Ad5-transformed cells. Further studies showed that the CTL is directed against the E1A region of Ad5 and not against the E1B region. The target peptide in the E1A protein for our CTL clone turned out to be a peptide of only eight amino acids, located close to the C-terminal end of the E1A protein: Pro Ser Asn Thr Pro Pro Glu Ile.

Since the bulk culture reacted with the same peptide as the individual CTL clones we assume that there is only one peptide recognized. The conclusions from this work [12] can be summarized as follows:

1. Complete regression of virally induced mouse tumours with sizes of up to 10 cm^3 was obtained, by a highly potent CTL clone in the presence of IL-2.
2. The lysis of the tumour cells is class I MHC-restricted.
3. Only one peptide in the Ad5 E1A protein is recognized (E1A proteins have a nuclear localization).
4. The analogous peptide is completely lacking in the related Ad12 E1A protein.
5. Natural killer cells do not play an important role in preventing adenovirus tumourigenesis under the conditions used.

Thus, the discoveries of oncogenes and tumour-suppressor genes over the past decade have provided for the first time insight into the molecular mechanisms of carcinogenesis. This knowledge may eventually contribute to the development of cancer therapy protocols, an example of which has been presented for an adenovirus tumour model.

References

1. Bishop JM (1987) The molecular genetics of cancer. Science 235:305–311
2. Klein G (1988) The approaching era of the tumor suppressor genes. Rev Oncol 1:427–437
3. Weinberg RA (1989) Oncogenes, antioncogenes, and the molecular bases of multistep carcinogenesis. Cancer Res 49:3713–3721
4. Bock G, Marsh J (eds) (1989) Genetic analysis of tumour suppression. John Wiley, New York, p 142 (Ciba Foundation Symposium)
5. Bos JL (1989) *Ras* oncogenes in human cancer: A review. Cancer Res 49: 4682–4689
6. Timmers HThM, Van Zoelen EJJ, Boss JL, van der Eb AJ (1988) Cells transformed by adenovirus type 12 but not by type 5 are dependent on insulin or insulin-like growth factors for their proliferation. J Biol Chem 263:1329–1335
7. Timmers HThM, De Wit D, Bos JL, van der Eb AJ (1988) E1A products of adenoviruses reduce the expression of cellular proliferation-associated genes. Oncogene Res. 3:67–76
8. Offringa R, Smits AMM, Houweling A, Bos JL, van der Eb AJ (1988) Similar effects of adenovirus E1A and glucocorticoid hormones on the expression of the metalloprotease stromelysin. Nucl Acids Res 16:10973–10984
9. Offringa R, Gebel S, Timmers HThM, Smits AMM, Zwart R, Cramer S, Stein B, Bos JL, van der Eb AJ, Herrlich P (1990) A novel function of the transforming domain of EIA: repression of AP-1 activity. Cell (in press)
10. Schrier PI, Bernards R, Vaessen RTMJ, Houweling A, van der Eb AJ (1983) Expression of class I major histocompatibility antigens switched off by highly oncogenic adenovirus 12 in transformed rat cells. Nature 305:771–775
11. Bernards R, Schrier PI, Houweling A, Bos JL, van der Eb AJ (1983) Tumorigenicity of cells transformed by adenovirus type 12 by evasion of T-cell immunity. Nature 305:776–779
12. Kast WM, Offringa R, Peters PJ, Voordouw AC, Meloen RH, van der Eb AJ, Melief CJM (1989) Eradication of adenovirus E1-induced tumors by E1A specific cytotoxic T lymphocytes. Cell 59:603–614

Chapter 7

Epstein–Barr Virus Infection and Cancer

M. A. Epstein

Introduction

Evidence linking Epstein–Barr (EB) virus to two human cancers, endemic Burkitt's lymphoma (BL) [1] and undifferentiated nasopharyngeal carcinoma (NPC) [2], has accumulated steadily over the years [3–5]. Already in 1976 enough was known for it to be evident that EB virus was an essential link in a complicated chain of events leading inexorably to these malignant tumours, and it therefore seemed reasonable to suppose that if infection by the virus could be prevented the incidence of the tumours in populations at risk would be decreased [6]. In view of the compelling analogy afforded by the effect of avoiding cigarette smoking on the incidence of bronchogenic carcinoma [7], it seemed unethical not at least to explore the possibility of vaccine intervention to prevent EB virus infection in the context of BL and NPC.

This concept has sometimes been criticized on the grounds that the mechanisms whereby EB virus might bring about malignant change were not known, nor even that the virus was in fact naturally carcinogenic. However, the mechanisms whereby cigarette smoking induces lung cancer are likewise not understood yet no-one would argue that cigarette smoking should not be discouraged until they are; and as regards EB virus and the causation of cancer in man, the only possible definitive proof in the human situation would come from the demonstration that a vaccine programme against the virus reduced the incidence of the associated malignant diseases. Accordingly a sustained effort has been made over the last dozen years to investigate the possibility of elaborating an antiviral vaccine to prevent infection with EB virus.

Rationale for an EB Virus Vaccine

Before the work was undertaken a number of questions required consideration. In the first place, would the effort to develop such a vaccine be worthwhile? Although BL is the most common cancer of children in endemic zones, more common there than all other children's cancers added together [1], in World cancer terms it is not numerically significant and where it is endemic there are far more pressing medical and public health problems to be resolved. NPC on the other hand is the most common cancer of men and the second most common cancer of women of southern Chinese origin and is thus the major cancer problem for huge populations in southern China and wherever southern Chinese have spread throughout South-East Asia and, indeed, the world [2]. In addition, there is an intermediate high incidence zone right across North Africa and down through the Sudan as far as the Kenya highlands [8,9]. Among the Inuit, NPC is also the most common cancer [10]. Thus, attempts at vaccine intervention would certainly be worthwhile.

The second question which had to be asked concerned possible precedents for this approach. Marek's disease of chickens was recognized as a curiosity many decades ago [11], but after World War II when huge commercial chicken flocks were established for battery egg-laying and the raising of broilers, infection with the herpesvirus of Marek's disease led to a steady 25% loss of birds from the malignant lymphomas induced by the agent [12]. This important economic problem was solved with the introduction of vaccines against Marek's herpesvirus [13,14] which virtually eliminated the disease and its lymphomas. This represented the first example of antiviral vaccine intervention leading to a reduction in the incidence of a naturally occurring cancer.

Although the vaccine in current use is an apathogenic live viral vaccine, experimental vaccines based on membranes from cells infected with the herpesvirus of Marek's disease or even virus-determined antigens purified from such membranes [15,16] proved highly effective. For this reason, and also because antibodies directed against the EB virus-determined membrane antigen (MA) were known to be virus-neutralizing [17], MA was chosen from the start as an appropriate vaccine immunogen.

Elaboration and Validation of a Prototype Subunit Vaccine

Work on the nature of MA from several laboratories showed that it consisted of two high molecular weight glycoprotein molecules of 270 and 340 kDa (MA gp270 and gp340) and further studies indicated that the former was a precursor of the latter (reviewed in [18]). Efforts were therefore made to develop a vaccine against EB virus based on MA gp340.

For this programme there were certain essential requirements. First, supplies of the only animal known to respond to EB virus infection with lesions had to be assured. The animal in question, the cottontop tamarin (*Saguinus oedipus oedipus*) [19], was placed on the endangered species list in the late 1970s and it was therefore necessary to establish a breeding colony. For this, the diet, bioenergetics, husbandry, and handling procedures for breeding had to be worked out [20,21] and it was also necessary to establish a challenge dose of EB virus which would ensure the induction of tumours in 100% of normal animals of this outbred population [22]. At the same time, a sensitive test for the gp340 antigen was essential to optimize yields of antigen and a radio-immunoassay (RIA) was therefore established [23]; with it an efficient MW-based preparative procedure was introduced [24]. The purified gp340 was incorporated into artificial liposomes to increase its immunogenicity. Finally, the antibodies induced were monitored by a specific enzyme-linked immunosorbent assay (ELISA) [25] and standard virus neutralization tests [17,26]. As soon as these requirements had been fulfilled, the prototype MA gp340-based vaccine was validated in cottontop tamarins by demonstrating its ability to protect vaccinated animals against the 100% carcinogenic dose of challenge EB virus [22].

Assessment of Vaccines for Use in Man

Over the last few years the work has concentrated on the development of a gp340-based vaccine suitable for human use. The gene coding for gp340 has been known for some time and has been sequenced [27,28], but although it has been cloned and expressed in bacterial [29], yeast [30] and mammalian cell systems [31,32] it has not proved possible to purify the product for use as a vaccine. Accordingly, other approaches have been followed up. Recombinant vaccinia viruses capable of expressing gp340 in tissue culture have been made [33] and have been tested in tamarins. The recombinant made from the rather virulent WR strain of vaccinia caused massive lesions when scarified on the skin (4–5 cm in diameter with a crop of satellite secondary pustules one week later) and the considerable amount of virus replication responsible for these lesions gave sufficient expression of the gp340 gene to ensure protection in some animals [34]. However, the WR strain of vaccinia virus is quite unsuitable for use in man and when the experiments were repeated with the Wyeth NY Board of Public Health vaccine strain of vaccinia the minimal virus replication of this attenuated strain gave insufficient expression of the gp340 gene to induce immunity [34].

A recombinant varicella virus has also been made using similar genetic engineering techniques to insert the gp340 gene [35], starting from the outset with the Oka vaccine strain of varicella [36]. This strain has been employed extensively to vaccinate young children and no adverse reactions have been reported. It has been shown recently that the common marmoset (*Calithrix jacchus*) can be infected experimentally with Oka varicella [37] and although this species does not respond with disease manifestations when infected with EB virus it is closely related to the cottontop tamarin which does. It may therefore be possible to investigate the protective effects of the recombinant Oka varicella against tumour induction by EB virus using tamarins and this hope has been strengthened by current experiments indicating that the recombinant will grow in cottontop tamarin fibroblast cultures (Morgan et al., unpublished work).

A First Generation MA gp340 Vaccine for Human Trials

In the meantime, more recent approaches have concentrated on new purification procedures for MA gp340 and the use of novel adjuvants. During a preliminary analysis of the structure of MA gp340 it was found that the carbohydrate moiety was heavily sialiated [38] and it became clear that the molecule should therefore remain negatively charged at a relatively low pH. Making use of this characteristic, a purification procedure was developed using anion exchange chromatography on an automated fast protein liquid chromatography (FPLC) system [39]. With this method the membrane fraction of EB virus-infected cells (expressing therefore MA gp340) is solubilized in a synthetic non-ionic detergent (MEGA-9) [40] and passed down an ion exchange matrix (Mono-Q) under conditions of low ionic strength at pH 5.0. The MA gp340 binds to the anion exchanger very efficiently and can be readily eluted with a salt gradient. Although virtual purity is achieved with this single step, a final gel filtration guarantees the homogeneity of the product [39]. The advantages of this procedure are that it isolates all MA gp340 molecules, avoids denaturing conditions, is both automated and reproducible, and has the potential for easy scale-up. With FPLC-purified MA gp340 readily available, its efficiency as a vaccine has been investigated in conjunction with two very efficient new adjuvants.

FPLC-purified gp340 has been incorporated in Quil A-derived immunostimulating complexes (iscoms) [41,42] which have powerful adjuvant properties. This combined material was highly immunogenic on inoculation into cottontop tamarins, inducing high titre virus-neutralizing antibodies and conferring protection against the standard 100% lymphomagenic dose of virus [43]. Although there are preliminary observations suggesting that iscoms are not significantly toxic locally [44], and they have already been used successfully in veterinary practice, a great deal more investigation will be required before they can be deemed suitable for use in man.

Accordingly, another new adjuvant preparation has been tested with FPLC-purified gp340. This is the synthetic threonyl analogue of muramyl dipeptide (MDP), which has the advantage of retaining the highly efficient adjuvant properties of MDP while being entirely free from the well known side effects of that compound [45,46]. Threonyl MDP emulsified in squalane together with a copolymer (L121) gives a suspension of microspheres on whose surface antigens are displayed and which activate complement. This adjuvant formulation is suitable for human use (Masterfile submitted to US Food and Drug Administration, 1988) and when tested with FPLC gp340 in tamarins rapidly induced high titre virus-neutralizing antibodies after three injections and conferred protection against tumour induction by the standard dose of challenge virus [47]. With this adjuvant formulation doses of antigen as low as 5 μg have proved effective.

It is clear from the foregoing that MA gp340 can now be easily prepared in tractable amounts by the FPLC purification procedure and that the MDP formulation provides a safe, easy to use, highly effective adjuvant giving excellent results with extremely small doses of antigen. A first generation subunit vaccine against EB virus is thus now available for trial in man.

Human Trials

The Phase I Trial

A Phase I human trial has been approved by two separate UK expert scientific committees and funds for such a trial have been allocated; it is proposed that

FPLC gp340 immunogen will be prepared on contract in the commercial sector under conditions of Good Manufacturing Practice. The objectives of the trial are as follows:

1. To determine the immunogenicity in man of FPLC-purified EB virus MA gp340 administered with the MDP adjuvant formulation by intramuscular injection.
2. To make a preliminary estimate of its tolerability including when given to EB virus seropositive individuals.

The protocol of the Phase I trial is planned in the following steps:

1. Serological screening for antibodies to EB virus of up to 100 young, informed, consenting human volunteers.
2. Selection of 12 seronegative individuals and 12 seropositive individuals with low levels of antibody to MA gp340.
3. Full clinical and laboratory screening of those chosen.
4. Full investigation of EB virus status of those chosen (serology, salivary virus shedding, establishment of B cell lines as subsequent targets for cell mediated immunity).
5. Administration of vaccine (three doses at two-weekly intervals).
6. Post immunization clinical monitoring and scrutiny of daily diary for possible symptoms.
7. Immunological assessment of antibodies and cell-mediated responses to EB virus.
8. Analysis of possible local or systemic reactions to the vaccine, assessment in seronegative individuals of induced antibody and cellular reactivity to gp340, and in seropositive individuals of a boost in these measures of immunity.

Priority Use of the Vaccine Against Duncan Syndrome

As soon as the Phase I trial proves satisfactory, and even before further studies of the vaccine, it would be ethically acceptable to use it in young boys belonging to families known to suffer from Duncan or X-linked lymphoproliferative syndrome [48], since in this condition primary EB virus infection of males is accompanied by life-threatening complications with a very high frequency of deaths.

Future Double Blind Trial

Should all go well in the Phase I trial it will be necessary to extend human studies to a larger scale double blind placebo trial. For this, a group of seronegative young adult volunteers will need to be investigated and will be randomly assigned to treatment with the vaccine preparation or an indistinguishable placebo. Thereafter the rate of seroconversion from natural infection will be monitored with the two groups noting whether the vaccine prevents seroconversion and decreases the incidence of infectious mononucleosis.

Future Field Trials

If this trial proves successful it will be necessary to assess the performance of the vaccine in a field trial in some area where the incidence of endemic BL is high. Since such regions are restricted to Africa and New Guinea [1] where primary EB virus infection occurs at a very early age [49,50] it will be necessary to vaccinate infants in the first few months after birth. The logistics for this are no more difficult than those required for similar vaccine programmes against hepatitis B virus infection in high incidence areas [51,52], and indeed the EB virus project could readily be linked to an ongoing WHO hepatitis B programme in the Gambia [53]. This small West African country has a number of special advantages for such work: (a) its geography is such that all of its 500 villages can readily be reached by road, (b) the population of some 700 000 is sufficient for the proposed vaccine studies, (c) the presence of a British Medical Research Council Tropical Medicine Research Unit over the last forty years has provided substantial baseline information on the health, social make up and practices of the inhabitants, and (d) for an EB virus vaccine trial, the existence of about 40 Burkitt's lymphomas per year is enough to monitor the ability of the vaccine to reduce the incidence of this cancer.

Final Comment

Once field studies have demonstrated that vaccination against EB virus infection can bring about a decrease in the incidence of BL in a high risk population, there will be an irresistible case for devoting funds and efforts to the development of more sophisticated vaccines against EB virus with the long-term aim of preventing primary infection in the huge populations for whom EB virus-associated NPC is the major cancer problem. Because this is a tumour of middle and later life [2] immunity to EB virus would probably need to be maintained for many decades and much study and effort will have to be devoted to the elaboration of methods to do this. Nevertheless, progress with a vaccine against EB virus has continued steadily and success at each step generates acceleration in the rate of advance.

References

1. Burkitt D (1963) A lymphoma syndrome in tropical Africa. In: Richter GW, Epstein MA (eds) International review of experimental pathology. Academic Press, New York London, pp 67–138
2. Shanmugaratnam K (1971) Studies on the aetiology of nasopharnygeal carcinoma. In: Richter GW, Epstein MA (eds) International review of experimental pathology, vol 10. Academic Press, New York London, p 361–413
3. Epstein MA, Morgan AJ (1983) Clinical consequences of Epstein–Barr virus infection and possible control by an antiviral vaccine. Clin Exp Immunol 53:257–271
4. Whittle HC, Brown J, Marsh K, Greenwood BM, Seidelin P, Tighe H, Wedderburn L (1984) T cell control of B cells infected with EB virus is lost during *P.falciparum* malaria. Nature 312:449–450

5. Lenoir GM, Bornkamm GW (1987) Burkitt's lymphoma, a human cancer model for the study of the multistep development of cancer; proposal for a new scenario. In: Klein G (ed) Advances in viral oncology series, vol 7. Raven Press, New York, pp 173–206
6. Epstein MA (1976) Epstein–Barr virus – is it time to develop a vaccine program? J Natl Cancer Inst 56:697–7000
7. Doll R, Peto R (1976) Mortality in relation to smoking: 20 years' observation on male British doctors. Br Med J ii:1525–1536
8. Cammoun M, Hoerner GV, Mourali N (1974) Tumors of the nasopharynx in Tunisia: an anatomic and clinical study based on 143 cases. Cancer 33:184–192
9. Clifford P (1970) A review: on the epidemiology of nasopharyngeal carcinoma. Int J Cancer 5:287–309
10. Lanier A, Bender T, Talbot M, Wilmeth S, Tschopp C, Henle W, Henle G, Ritter D, Terasaki P (1980) Nasopharyngeal carcinoma in Alaskan Eskimos, Indians and Aleuts: a review of cases and study of Epstein–Barr virus, HLA and environmental risk factors. Cancer 46:2100–2106
11. Marek J (1907) Multiple Nervenentzündung (polyneuritis) bei Hühnern. Dt-Öst Tierärztl Wschr 15:412–421
12. Payne LN, Frazier JA, Powell PC (1976) Pathogenesis of Marek's disease. In: Richter GW, Epstein MA (eds) International review of experimental pathology, vol 16. Academic Press, New York, San Francisco London. pp 59–154
13. Churchill AE, Payne LN, Chubb RC (1969) Immunization against Marek's disease using a live attenuated virus. Nature 221:744–747
14. Okazaki W, Purchase HG, Burmester BR (1970) Protection against Marek's disease by vaccination with a herpesvirus of turkeys. Avian Dis 14:413–429
15. Kaaden OR, Dietzschold B (1974) Alterations of the immunological specificity of plasma membranes of cells infected with Marek's disease and turkey herpes viruses. J Gen Virol 25:1–10
16. Lesnick F, Ross LJN (1975) Immunization against Marek's disease using Marek's disease virus-specific antigens free from infectious virus. Int J Cancer 16: 153–163
17. De Schryver A, Klein G, Hewetson J, Rocchi G, Henle W, Henle G, Moss DJ, Pope JH (1974) Comparison of EBV neutralization tests based on abortive infection or transformation of lymphoid cells and their relation to membrane reactive antibodies (anti MA). Int J Cancer 13:353–362
18. Epstein MA (1984) A prototype vaccine to prevent Epstein–Barr (EB) virus-associated tumours. Proc R Soc Lond B 221:1–20
19. Miller G, Shope T, Coope D, Waters C, Pagano J, Bornkamm GW, Henle W (1977) Lymphoma in cotton-top marmosets after inoculation with Epstein–Barr virus: tumor incidence, histologic spectrum, antibody responses, demonstration of viral DNA, and characterization of viruses. J Exp Med 145:948–967
20. Kirkwood JK, Epstein MA, Terlecki AJ (1983) Factors influencing population growth of a colony of cotton-top tamarins. Lab Anim 17:45–41
21. Kirkwood JK, Epstein MA, Terlecki AJ, Underwood SJ (1985) Rearing of a second generation of cotton-top tamarins (*Saguinus oedipus oedipus*) in captivity. Lab Anim 19:269–272
22. Epstein MA, Morgan AJ, Finerty S, Randle BJ, Kirkwood JK (1985) Protection of cottontop tamarins against Epstein–Barr virus-induced malignant lymphoma by a prototype subunit vaccine. Nature 318:287–289
23. North JR, Morgan AJ, Thompson JL, Epstein MA (1982) Quantification of an EB virus-associated membrane antigen (MA) component. J Virol Meth 5:55–65
24. Morgan AJ, North JR, Epstein MA (1983) Purification and properties of the gp340 component of Epstein–Barr (EB) virus membrane antigen (MA) in an immunogenic form. J Gen Virol 64:455–460
25. Randle BJ, Epstein MA (1984) A highly sensitive enzyme-linked immunosorbent assay to quantitate antibodies to Epstein–Barr virus membrane antigen gp340. J Virol Meth 9:201–208
26. Moss DJ, Pope JH (1972) Assay of the infectivity of Epstein–Barr virus by transformation of human leucocytes *in vitro*. J Gen Virol 17:233–236
27. Hummel M, Thorley-Lawson DA, Kieff E (1984) An Epstein–Barr virus DNA fragment encodes messages for the two major envelope glycoproteins (gp350/300 and gp220/200). J Virol 49:413–417
28. Biggin M, Farrell PJ, Barrell BG (1984) Transcription and DNA sequence of *Bam* HIL fragment of B95-8 Epstein–Barr virus. EMBO J 3:1083–1090
29. Beisel C, Tanner J, Matsuo T, Thorley-Lawson D, Kezdy F, Keiff E (1985) Two major outer envelope glycoproteins of Epstein–Barr virus encoded by the same gene. J Virol 54:665–674
30. Schultz LD, Tanner J, Hofmann K, Emini E, Kieff E, Ellis RW (1987) Expression and analysis of

EBV gp350 in yeast *Saccharomyces cerevisiae*. In: Levine PH, Ablashi DV, Nonoyama M, Pearson GR, Glaser R (eds) Epstein–Barr virus and human disease. Humana Press, Clifton, New Jersey, pp 475–478
31. Whang Y, Silberklang M, Morgan A, Munshi S, Lenny AB, Ellis RW, Kieff E (1987) Expression of Epstein–Barr virus gp350–220 gene in rodent and primate cells. J Virol 61:1796–1807
32. Conway M, Morgan A, Mackett M (1989) Expression of Epstein–Barr virus antigen gp340/220 in mouse fibroblasts using a bovine popilloma virus vector. J Gen Virol 70:729–734
33. Mackett M, Arrand JR (1985) Recombinant vaccinia virus induces neutralizing antibodies in rabbits against Epstein–Barr virus membrane antigen gp340. EMBO J 4:3229–3234
34. Morgan AJ, Mackett M, Finerty S, Arrand J, Scullion F, Epstein MA (1988) Recombinant vaccinia virus expressing Epstein–Barr virus glycoprotein gp340 protects cottontop tamarins against EB virus-induced lymphomas. J Med Virol 25:189–195
35. Lowe RS, Keller PM, Keech BJ, Davison AJ, Whang Y, Morgan AJ, Kieff E, Ellis RW (1987) Varicella-zoster virus as a live vector for the expression of foreign genes. Proc Natl Acad Sci USA 84:3896–3900
36. Takahaski M, Otsuka T, Okuno Y, Asano Y, Yazaki T, Isomura S (1974) Live varicella vaccine used to prevent the spread of varicella in children in hospital. Lancet ii:1288–1290
37. Provost PJ, Keller PM, Banker FS, Keech BJ, Klein HJ, Lowe RS, Morton DH, Phelps AH, McAleer WJ, Ellis RW (1987) Successful infection of the common marmoset (*Callithrix jacchus*) with human varicella-zoster virus. J Virol 61:2951–2955
38. Morgan AJ, Smith AR, Barker RN, Epstein MA (1984) A structural investigation of the Epstein–Barr (EB) virus membrane antigen glycoprotein, gp340. J Gen Virol 65:397–404
39. David EM, Morgan AJ (1988) Efficient purification of Epstein–Barr virus membrane antigen gp340 by fast protein liquid chromatography. J Immunol Meth 108:231–236
40. Hildreth JEK (1982) *N*-D-Gluco-*N*-methylalkanamide compounds, a new class of non-ionic detergents for membrane biochemistry. Biochem J 207:363–366
41. Morein B, Sundquist B, Höglund S,Dalsgaard K, Osterhaus A (1984) Iscom, a novel structure for antigenic presentation of membrane proteins from enveloped viruses. Nature 308:457–460
42. Morein B, Lövgren K, Höglund S, Sundquist B (1987) The iscom: an immunostimulating complex. Immunol Today 8:333–338
43. Morgan AJ, Finerty S, Lovgren K, Scullion FT, Morein B (1988) Prevention of Epstein–Barr (EB) virus-induced lymphoma in cottontop tamarins by vaccination with the EB virus envelope glycoprotein gp340 incorporated into immune-stimulating complexes J Gen Virol 69:2093–2096
44. Speijers G, Danse L, Bewary E, Strik J, Vos J (1988) Local reactions of the saponin Quil A and a Quil A-containing iscom measles vaccine after intramuscular injection of rats: a comparison with the effects of DPT-polio vaccine. Fundamental Appl Toxicol 10:425–430
45. Allison AC, Byars NE (1986) An adjuvant formulation that selectively elicits the formation of antibodies of protective isotype and cell mediated immunity. J Immunol Meth 95:157–168
46. Allison AC, Byars NE (1987) Vaccine technology: adjuvants for increased efficiency. Biotechnology 5:1041–1045
47. Morgan AJ, Allison AC, Finerty S, Scullion FT, Byars NE, Epstein MA (1989) Validation of a first generation Epstein–Barr virus vaccine preparation suitable for human use. J Med Virol 29:74–88
48. Purtilo DT, Sakamoto K, Barnabei V, Seeley J, Beechtold T, Rogers G, Yetz J, Harada S (1982) Epstein–Barr virus-induced diseases in boys with the X-linked lymphoproliferative syndrome (XLP). Am J Med 73:49–56
49. Henle W, Henle G (1969) The relation between the Epstein–Barr virus and infectious mononucleosis, Burkitt's lymphoma and cancer of the postnasal space. E Afr Med J 46:402–406
50. De-Thé G (1979) Demographic studies implicating the virus in the causation of Burkitt's lymphoma; prospects for nasopharyngeal carcinoma. In: Epstein MA, Achong BG (eds) The Epstein–Barr virus. Springer, Berlin Heidelberg New York, pp 417–443
51. Zuckerman AJ (1985) Prevention of hepatocellular carcinoma by immunization against hepatitis B. In: Epstein MA, Richter GW (eds) International review of experimental pathology, vol 25. Academic Press, Orlando, pp 59–81
52. Deinhardt F, Jilg W (1986) Vaccines against hepatitis. Annls Inst Pasteur, Virol 137E:79–95
53. International Agency for Research on Cancer (1985) An intervention study to evaluate the effectiveness of hepatitis B vaccine for the prevention of hepatocellular carcinoma in a high risk population. IARC Working Paper 3/6:1–46

Section III

MECHANISMS AND THEIR IMPLICATIONS FOR PREVENTION AND THERAPY

Chapter 8

Chronic Infection, Immunopathology and Immune Suppression

N. A. Mitchison

Introduction

As an exercise in evolutionary theory, this paper will excite criticism on all sides: from experimentalists as too speculative, from theoreticians as unprofessional, and from the young as evidence of a decline in the faculties. Well, yes, but is it not one of the Rights of Man to formulate ideas, erratic or not?

My theme is the loss of polymorphic genes, and the remarkable things that can happen to them on the way out. We shall be concerned chiefly with the major histocompatibility complex (MHC), but will touch also on V genes where the process of loss can be discussed in a rather simpler way. Together, these two gene pools represent the most polymorphic part of the immune system, and probably also of the human genome. Both these polymorphisms are driven by what evolutionary theorists have come to call the Red Queen strategy, after the character in Lewis Carroll's *Through the Looking Glass* who keeps running but stays in the same place. This mechanism is believed to drive much of evolution in general, for any one species exists in a changing environment of other species, and, as they evolve, so that species comes under selective pressure itself to evolve so as to keep up. We should recall Haldane's rule, that most of the cost of evolution is paid for the maintenance of polymorphism, and can then see that the place where the Red Queen strategy applies *par excellence* is in relation to these two polymorphisms of the immune system, as driven by parasites. In both polymorphisms, parasites (using this term in the extended sense to include everything from viruses to worms) come under selective pressure to vary, so as to escape detection by these immunological recognition molecules; and the recognition molecules vary so as to prevent the escape. Heterozygous hosts have a selective advantage because they are able to recognize a wider variety of parasite antigens, and this is what maintains the polymorphisms in balance.

In returning to this theme (see also [1–4]) it is a pleasure to acknowledge the guidance that I have received from David Oliveira, Takehiko Sasazuki, René de Vries and Richard Dawkins, and the important work of Richard Parham [5].

V-TcR Gene Entry and Exit

Entry of new V genes is presumably driven by the Red Queen strategy, as set out above. Exit, and the factors that set an upper limit to the total number of these genes are less well understood. A full discussion of them would take us into the deep waters of DNA recombination and replication mechanisms. However there is one recently identified simple exit pressure, that provides an excellent illustration of the way in which these pressures operate and also lets me introduce the elegant Fig. 8.1

This figure indicates some of the ways in which parasites exploit and manipulate molecules of the immune system. References are as follows: rhinovirus [6]; bacterial exotoxins [7]; human immunodeficiency virus [8]; adenovirus [9]; Epstein–Barr virus [10]. Exploitation of this sort will select for counter-measures on the part of the host. One simple but drastic measure is illustrated in Fig. 8.2. Kappler's proposal is that some mice simply delete the parts of their collection of V-TcR genes that encode toxin-binding receptors. This is in line with the fact that wild mice are highly polymorphic for deletions of sets of these genes.

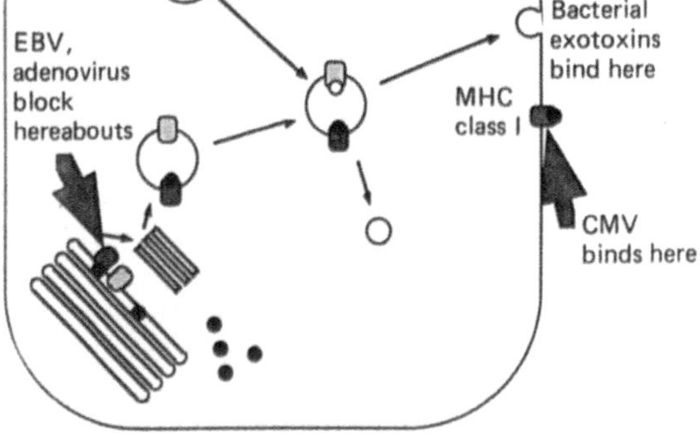

Fig. 8.1. Interaction with pathogens exerts evolutionary pressure for immunological molecules to become "slippery" or disappear.

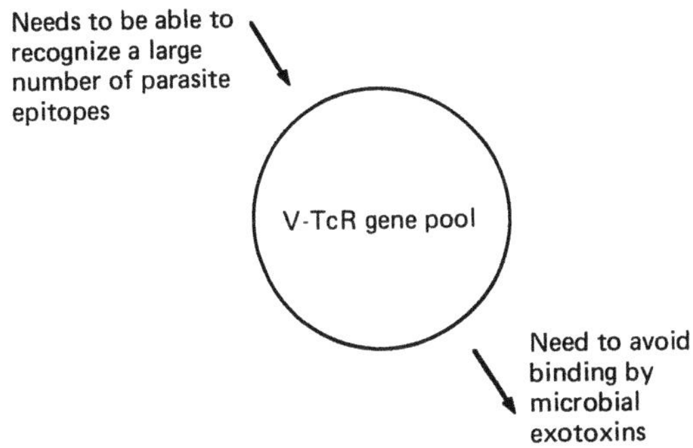

Fig. 8.2. Evolutionary pressures on the V-TcR gene pool (after J. Kappler.)

As there are many different toxins, each probably binding to different sets of TcRs, the polymorphism must be fairly complex.

MHC Gene Entry and Exit

Again entry is driven by the Red Queen strategy. The main pressure driving exit can be identified as the need to avoid punching too many holes in the T cell receptor repertoire. Each MHC gene expressed in an individual will punch its own hole, by means of clonal selection within the thymus, where these genes operate a mechanism of negative selection. So do many other genes, but the holes created by those of the MHC are unusually large; in the Marrack–Kappler terminology the MHC molecules are super-antigens, able to produce drastic effects because they bind many different peptides derived from self-proteins.

From the parasite's point of view, there are two avenues of escape from the immune response. One is by varying the structure of its antigens so as to delete sequences recognizable by MHC molecules. The other, so to speak, is to duck into the holes created in the repertoire by the MHC. The latter process is none other than the familiar concept of antigenic mimicry, here used in the limited sense of mimicry of the MHC.

This drive towards exit can be seen to have a number of consequences. One will be a reduction in polymorphism, that is, in the likelihood of an individual carrying two different alleles at an MHC locus. Clearly a homozygous gene will punch a smaller hole in the repertoire than one which is heterozygous (by a factor of slightly less than a half, because the holes punched by two alleles are likely to overlap somewhat). Another will be a reduction in the total number of MHC genes. It is notable that although the number of MHC genes varies from one mammalian species to another, and these genes enter and leave the genome from time to time during the course of evolution, the variation is confined within fairly narrow limits. In contrast, the number of V genes on any one chromosome can vary enormously. A third consequence will be reduced expression, as an MHC gene with reduced expression in the thymus will have a diminished impact on the T cell repertoire. A fourth possibility is that the species may become polymorphic for non-expression of an MHC gene, as has occurred for the H-2 E locus of mice, although it is not clear whether this is a common occurrence in other species or exactly how it fits into the present argument.

Figure 8.3. illustrates the composition of the MHC from this point of view, in what I trust is not an over-dramatized way. All MHCs that have been adequately studied turn out to be littered with genes showing reduced expression, and with pseudogenes. The littering gives the impression of having occurred progressively, with reduced expression as a stage on the road to complete loss. That, of course, does not imply that mutations are more likely to occur in control sequences than in coding ones. It is rather a matter of selection, as a gene once demoted to pseudogene status becomes likely to accumulate further mutations rapidly.

We come now to the nub of the argument. Is it possible that a gene on the way to exit, with reduced polymorphism and reduced expression as suggested above, may tend to acquire special functional properties? I believe that a logical argument can be constructed in favour of this contention, along the lines illustrated in Fig. 8.4. The idea presented there is that such a gene would come under selective pressure to mediate dominantly inherited immunological unresponsiveness (DIIU) This property might be implemented by various mechanisms, of which the two most obvious are suppressor T cells and negative selection in the thymus discussed further below. The argument runs as follows. Once an MHC gene begins to mediate unresponsiveness the pressure to vary diminishes: the Red Queen comes out of play. From a parasite's point of view such a gene is obviously advantageous, as it prevents something that would otherwise destroy the parasite, and so the pressure to vary antigens recognized by its product disappears. From the point of view of the host there could also be an advantage, in that unresponsiveness might protect against damage that would otherwise be caused by the response. As Dawkins would put it, in these circumstances

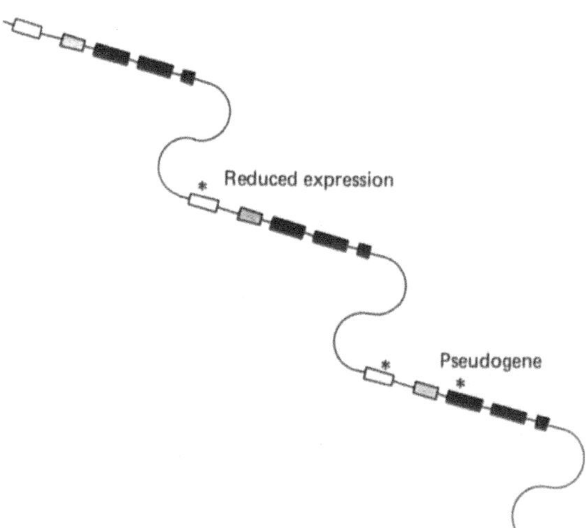

Fig. 8.3. The MHC as a mess.

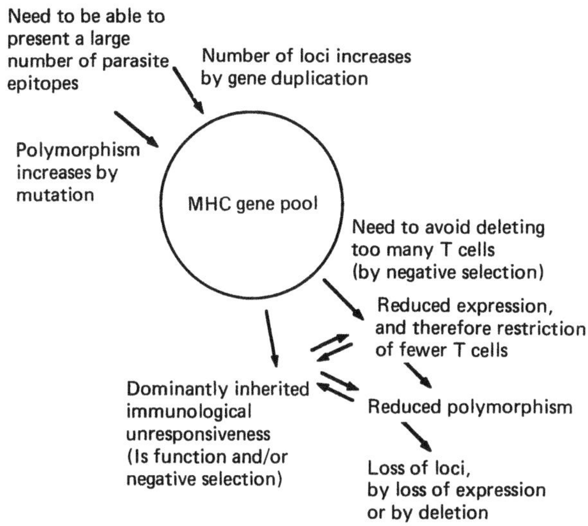

Fig. 8.4. Evolutionary pressures on the MHC gene pool.

immunological unresponsiveness would become an item on the joint desiderata list of host and parasite.

Damage of this sort is categorized as immunopathological, and the reactions that cause it are termed hypersensitivity. Figure 8.5. illustrates this situation as it applies to a "polar" disease, such as leprosy, but other infectious diseases also have polar features of this sort. In malaria, for instance, splenomegaly and interstitial renal damage, both involving hypersensitivity, are important sources of morbidity, and the same is true of elephantiasis in filariasis. In fact the situation depicted in Fig. 8.5. seems to be common to most chronic infections.

That is not the end of the story, for the argument from population genetics does not specify a mechanism of DIIU. From the genetic angle this is not a problem, as the evolution of dominance in general is well understood in terms of selection of modifiers and allele substitution, and the details of mechanism are of little concern. What matters is that all MHC genes are in principle subject to these

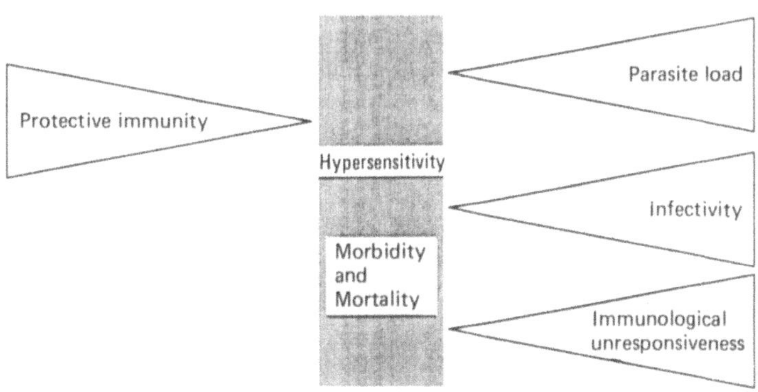

Fig. 8.5. Hypersensitivity develops in the intermediate zone between protection and suppression.

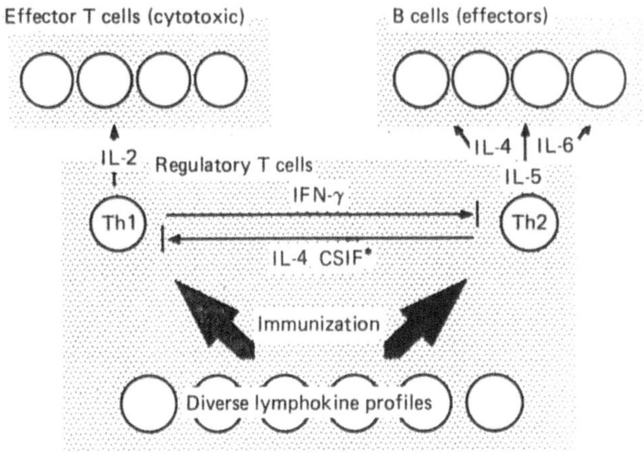

Fig. 8.6. Mutual inhibition by Th1 and Th2 cells explains some suppressor phenomena (after M. Berton, T. R. Mossmann, T. J. Powell and J. W. Streilein). Cytokine secretion inhibition factor (*CSIF*) has been cloned by T. R. Mossmann and co-workers, 1989.

evolutionary pressures, and that somehow or other they come to focus on particular genes. The feedback between these pressures is poorly understood (by me at least), hence the bi-directionality of the arrows connecting DIIU with reduced expression and reduced polymorphism in Fig. 8.4.

For the immunologist, on the other hand, it is precisely these details of mechanism that are of greatest interest. The key may be a class II MHC gene with reduced expression which thereby signals preferentially to suppressor T cells. This is an area where one is free to speculate, as so little is known for sure about mechanisms of suppression. In a recent review I have drawn attention to the division of opinion on this subject, between the strict school which continues to argue in favour of them as unique cells operating a unique molecular mechanism, and the assimilationist school which regards them as conventional T cells operating in a special context [11]. My own view is that the assimilationists are having the better of it. Their point of view, or one of their points of view, is expressed in Fig. 8.6 which presents Mossmann's splitting of the CD4 T cell subset, based on his presentation at the 1989 International Congress of Immunology. One could easily imagine that reduced MHC expression could enhance the mutual suppression of Th1 and Th2 cells.

Note here a complication, that there is no obvious way of distinguishing between genes entering and exiting from the gene pool. This reflects the general difficulty in distinguishing whether an evolutionary form is "primitive" or "degenerate", and will come as no surprise to any student of evolution.

HLA-DQ and H-2E

At the end of this perhaps over-theoretical discussion let us review briefly information concerning the two MHC genes that are the most likely candidates for the role outlined above, namely HLA-DQ in man and H-2 E in mouse. Such a review is presented in Tables 8.1 and 8.2; it deals with their function in mediating DIIU, but for reasons of space does not cover the polymorphism or level of expression of these genes. The information contained in the tables indicates that these two genes do indeed tend to mediate DIIU, although this is evidently a bias rather than an exclusive commitment, just as the foregoing argument would predict. What is much less clear is how DIIU operates, whether by suppressor T cells, MHC mimicry by parasites, or by some other mechanism. That issue is summarized for H-2 E in Fig. 8.7, which is slightly modified from the one presented in [4], and which suggests the kind of evidence needed to resolve the matter.

Table 8.1. Dominantly inherited immunological unresponsiveness (DIIU), its control by the MHC, and antigen-specific immune suppression: in man

1.	DIIU offers a convincing explanation of the familial distribution of at least one immunological disease, Cryptomeria pollenosis [12]
2.	In contrast, certain other forms of pollenosis conform to a pattern of dominantly inherited immunological responsiveness [13]
3.	DIIU under HLA-DQ control also explains the familial distribution of at least one infectious disease, lepromatous leprosy [14]
4.	DIIU under HLA-DQ control explains the population distribution of at least one antibody response to a virus, hepatitis B [15]
5.	DIIU under HLA-DQ control explains population distribution of resistance to IDDM [16]
6.	Depletion of CD8 T cells provides direct evidence of suppression in lepromatous leprosy [17], Cryptomeria pollenosis and other diseases [15]
7.	Relief of inhibition of proliferation by means of monoclonal antibody implicates HLA-DQ as restriction element for suppressor–inducer T cells in leprosy, schistosomiasis, hepatitis and pollenosis [15,18]
8.	In contrast, certain suppressor clones from borderline leprosy are restricted by HLA-DR (R. R. P. de Vries, personal communication)

Table 8.2. DIIU, its control by the MHC, and antigen-specific immune suppression: in mouse

1.	Several immunological responses show a DIIU pattern in the B10.A-B10 recombinants (4R, 5R) that implies control of unresponsiveness by H-2 E [2,19]
2.	In some but not all of these instances direct evidence of Is function of H-2 E has been obtained, e.g. by relief of inhibition of proliferation by means of monoclonal antibody in the case of F-liver protein-specific suppression [20]
3.	In other instances a search for suppression has yielded negative results, for example with the transgenic H-2E gene which confers protection against autoimmune insulitis (T. Kishimoto, personal communication), and there is now direct evidence for negative selection in this context [21]
4.	The H-2 E molecule is now known to be able to delete a substantial part of the T cell repertoire within the thymus, including Vβ8, and this provides an explanation of DIIU alternative to suppression (see also Fig. 8.6)
5.	A second DIIU pattern has now been recorded in the B10.A × B10 recombinant strains, under the control of H-2A^{b}. This can be explained in terms of suppressor function flipping from E to A in a haplotype where E is not expressed (unpublished data)

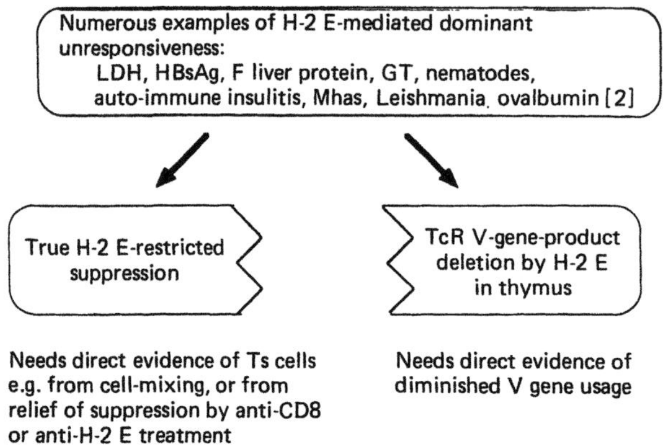

Fig. 8.7. Alternative mechanisms able to mediate dominantly inherited immunological unresponsiveness.

It is of course a mistake to put HLA-DQ and H-2 E on exactly the same footing. One notable difference is the high frequency of haplotypes that do not express H-2 E, alluded to above. This makes it likely that a relatively high proportion of the DIIU controlled by this gene will result from negative selection in the thymus.

Let me make one final point that bears on the issue. Before preparing this chapter I was obsessed with this question of the mechanism of DIIU, and it seemed to be of concern also to others [16,21]. I have now come to realize that in evolutionary terms this is a secondary issue, and that a logical structure connecting DIIU with other properties of an MHC gene exists independently of the precise mechanism by which DIIU operates. This seems to me an important message, particularly so for immunologists concerned with hypersensitivity and suppression in chronic infectious disease.

References

1. Mitchison NA (1981) Information transfer between the minor antigen and T cell receptor repertoires. *Scand J Immunol* 14:631–635
2. Oliveira DBG, Mitchison NA (1989) Immune suppression genes. Clin Exp Immunol 75:167–177
3. Mitchison NA (1989) Immunity to parasites: the evolutionary biology of parasitism. In: Parasitology BSP Autumn Symposium, Suppl.
4. Mitchison NA (1989) Is genes in the mouse. In: Melchers F et al. (eds) Progress in Immunology VII. Springer, Berlin, pp 845–852
5. Parham P, Benjamin RJ, Chen BP, Clayberger PD, Ennis AM, Krensky DA, Lawlor DA, Littman DR, Norment AM, Salter RD, Zemmour J (1989) Polymorphism peptide binding and CD8 interactions of class I MHC molecules. In: Immunological Recognition, LIV Cold Spring Harbor Symposium on Quantitative Biology, 1989, pp 529–543
6. Greve JM, Davis G, Meyer AM, Forte CP, Yost SC, Marlor CW, Kamarck ME, McClelland A (1989) The major human rhinovirus receptor is ICAM-1. Cell 59:839–848
7. Kappler J, Pullen A, Herman A, Choi Y, Callahan J, White J, Kushnir E, Gelfand E, Herron L, Kotzin B, Wakeland E, Potts W, Marrack P (1989) Vβ control of the response of T cells to

superantigens. In: Immunological Recognition, LIV Cold Spring Harbor Symposium on Quantitative Biology, 1989, pp 401–407
8. Sattentau QJ, Dalgleish AG, Weiss RA, Beverley PCL (1986) Epitopes of the CD4 antigen and HIV infection. Science 234:1120–1123
9. Burgert HG, Kvist S (1987) The E3/19K protein of adenovirus type 2 binds to the domains of the histocompatibility antigen required for CTL recognition. EMBO J 6:2019–2026
10. Townsend A, Ohlen C, Foster L, Bastin J, Karre K (1989) A mutant cell in which association of class I heavy and light chains is induced by viral peptides. In: Immunological Recognition, LIV Cold Spring Harbor Symposium on Quantitative Biology, 1989, pp 299–308
11. Mitchison NA (1989) Suppression. Immunol Today 10:392–393
12. Sasazuki T, Nishimura Y, Muto M, Saito Y (1984) MHC-linked immune suppression genes and their role in immunological disorders. In: Sasazuki T, Tada T (eds) Immunogenetics: its application to clinical medicine. Academic Press, New York, pp 21
13. Marsh DG, Zwollo P, Huang SK, Ghosh B, Ansari AA (1989) Molecular studies of human responses to allergens. In: Immunological Recognition, LIV Cold Spring Harbor Symposium on Quantitative Biology, 1989, pp 459–470
14. Ottenhoff, THM, Gonzalez NM, de Vries, RRP, Convit J, van Rood JJ (1984) Association of HLA specificity LB-E12 (MC1, DC1, MT1) with lepromatous leprosy in a Venezuelan population. Tissue Antigens 24:25–29
15. Sasazuki T (1989) Differential roles of HLA-DR and DQ molecules in regulation of immune response. In: Melchers F et al. (eds) Progress in Immunology VII. Springer, Berlin, pp 853–860
16. Todd JA, Bell JI, McDevitt HQ (1987) HLA-DQ gene contributes to susceptibility and resistance to insulin-dependent diabetes mellitus. Nature 329:599–604
17. Mehra V, Brennan PJ, Rada E, Convit J, Bloom BR (1984) Lymphocyte suppression in leprosy induced by *M.leprae* glycolipids. Nature 308:194–196
18. Ottenhoff T (1989) Antigen specific T cell nonresponsiveness in man: leprosy, an immunological model disease. In: Melchers F et al. (eds) Progress in Immunology VII. Springer, Berlin, pp 890–893
19. Baxevanis CN, Nagy ZA, Klein J (1981) A novel type of T-T cell interaction removes the requirement for I-B region in the H-2 complex. Proc Natl Acad Sci USA 78:3809–3813
20. Oliveira DBG, Blackwell N, Virchis AE, Axelrod RA (1985) T helper and T suppressor cells are restricted by the A and E molecules, respectively, in the F antigen system. Immunogenetics 22:169–175
21. Reich E-P, Sherwin RA, Kanagawa O, Janeway CA Jr (1989) An explanation for the protective effect of the MHC class II I-E molecule in murine diabetes. Nature 341:326–328

Chapter 9

What Drives Autoantibodies? The Evidence From Spontaneous Human Autoimmune Diseases

P. H. Plotz

Introduction

Although the fathers of immunology recognized from the beginning the danger that a competent immune system posed to its possessor, it was decades before an actual disease was connected to an antibody directed at a self-antigen. Soon, however, other antibodies directed at self components turned up in diseases, but their connection to the disease was far less clear than in Coombs-positive haemolytic anaemia. In fact, autoantibodies have become, especially in the rheumatological autoimmune diseases, far more numerous than plausible explanations of how they got there and what they are doing.

Immunology and rheumatology were rapidly intertwined. Rheumatoid factor and the LE cell conveniently appeared in two diseases that had remained obscure in the light cast by the germ theory of disease, and the family changed its name from "collagen" to "autoimmune". What brought these diseases together was, and still is, that they are chronic inflammatory illnesses from which no micro-organisms can be regularly and reliably isolated and in some of whose victims autoantibodies can be found. Although the pathological picture in the affected tissues of all of them is dominated by infiltrating T lymphocytes, most energy has been devoted to explaining the pathogenetic role of the autoantibodies.

The recognition that the majority of autoantibodies in human diseases are directed against intracellular components, however, and that they are, therefore, unlikely agents of pathogenesis, has moved the focus away from pathogenesis. Now, it is expected that understanding the origin of autoantibodies will shed light upon the aetiology of the diseases in which they are found, a view which I share.

Immunologists have spent a century observing what happens when nature – or her surrogate, the experimental immunologist – presents a foreign molecule to a creature. Complicated events unfold, and we study them with the tools of the day. We studied the antibodies, the T cells, the suppressor cells, and now we study V selection, joining, rearrangement, allelic exclusion, class-switching, somatic mutation, and their T cell equivalents. By contrast, the autoantibodies in the serum of a patient are the end product of that series of events and we must track back to the beginning: What molecules and what series of events *could* have resulted in the antibodies? At the outset, all we see clearly is the antibodies themselves and we want to know their origin. I will summarize where I think we stand today, what the holes in our knowledge are, and what kinds of experiments are most likely to yield relevant answers.

Let me begin by a diversion on the question of tolerance because a central issue in the analysis of autoimmunity has been whether or not it somehow violates tolerance. A great deal is known about T cell tolerance and its mechanisms, about the central and peripheral elimination of self-reactive T cells, above positive and negative selection, about the shape of T cell receptors and their interaction with major histocompatibility complex (MHC) products containing antigen fragments, and even about the ways in which T cells are eliminated [1]. There are, for T cells, a well-defined time and place for the elimination of forbidden clones. Furthermore, because the T cell receptor does not change by somatic mutation, long-lasting tolerance is achievable. By contrast, although a great deal is known about immunoglobulin structure, less is known about the critical signals for B cell differentiation [1]. By the time an antibody has been secreted and is available for study, after all, its cell of origin is beyond control by any antigen-specific mechanism. Although older experiments, now supported by transgenic models, demonstrate that B cells may be silenced [2] or eliminated [3], the lifelong ontogeny of the B cell compartment should allow escape. B cells, it seems to me, are not unlike Victor Laszlo in *Casablanca*. You remember how Ingrid Bergman persuades the evil Nazi, Major Strasser, that to eliminate her brave husband, a leader in the French underground stranded in Vichy-controlled Morocco, will not solve the problem of the resistance. "If you kill Victor Laszlo," she breathes, "a hundred Victor Laszlos will rise up to replace him." Thus, it seems, continuing V_H selection, junctional diversity and, above all, somatic mutation will permit escape from B cell tolerance unless B cell silencing and elimination is as wastefully active in the periphery throughout the lifetime of an animal as negative T cell selection is transiently during T cell ontogeny. It is possible that new observations on the local events in lymph nodes, especially in germinal centres, will change this picture.

The Character of the Autoantibodies

It is by now a cliché that the autoantibodies associated with autoimmune disease are "polyclonal", "antigen-driven", and often "T cell-dependent". If one puts aside the data drawn from Epstein–Barr virus (EBV)-transformed or hybridoma fusion lines from patients with autoimmune diseases on the grounds that the immortalized cells are unrepresentative of those producing disease-associated autoantibodies, the data justifying this generalization are roughly as follows.

Disease-associated autoantibodies are generally isotype-switched. Thus, in both the MRL mouse with lupus-like illness [4,5] and the patient with systemic lupus erythematosus [6], antibodies of one or several IgG subclasses are found. The same is true of anti-Ro antibodies [7], anti-La antibodies [8], anti-thyroid antibodies [9.10], anti-ribosomal antibodies [11], the antibodies of pemphigus [12], anti-Sm antibodies [13], anti-acetylcholine receptor antibodies [14], and even of the rheumatoid factor associated with rheumatoid arthritis [15]. As is characteristic for many immune responses, one or another subclass may predominate. In a recently completed study, we have examined the subclass distribution of the autoantibodies characteristic of a certain group of patients with idiopathic inflammatory myopathy (polymyositis or dermatomyositis). These antibodies, called anti-Jo-1, are directed at the enzyme histidyl-tRNA synthetase and are the most common example of antibodies directed against aminoacyl-tRNA synthetases or other translation-related cytoplasmic factors associated with this disease. We have found that although all patients have IgG1 anti-Jo-1 antibodies, and that these are almost always quantitatively the most important, some patients have IgG2, IgG3, IgA, or IgM, as well [16].

In itself, isotype switch is insufficient proof that the antibodies are from an antigen-driven secondary response. In early primary responses, isotype switch can occur in cells which do not turn on the somatic mutation mechanism and therefore do not bear the hallmarks of antigen-driving [17]. More potent evidence comes from the fact that autoantibodies show somatic mutation. Here, the most persuasive evidence by far is from hybridomas prepared from individual mice with a naturally-occurring autoimmune illness. In MRL mice, the rheumatoid factor from individual mice is clearly derived from cells that have undergone somatic mutation [18]. It is possible to construct family trees of the hybridomas, demonstrating not only that the lines are related, but that they almost certainly arose from a single clone rather than multiple clones. In this instance, then, the claim is for somatic mutation rather than polyclonality. Individual mice, however, though genetically inbred and presumably identical with respect to their V region repertoires, use different clones. In NZB/NZW mice with lupus, a similar tale can be told for the hybridomas with anti-DNA activity [4,5,19].

In human autoimmune diseases, because no similar family trees have yet been constructed, the evidence for somatic mutation must rest upon other grounds. For several autoantibodies, it has been possible to show that there is a broad spectrotype of the antibodies. This has been shown for anti-DNA antibodies, for anti-Sm antibodies [20], for anti-thyroid antibodies [21], and for anti-Jo-1 antibodies [16]. In our studies of anti-Jo-1 antibodies, we have shown not only that at a single point in time the antibodies have the broad spectrotype of the typical secondary response, but also that in an individual patient, the spectrotype is stable despite fluctuations in the amount of antibody. This stability of spectrotype is found also in antithyroglobulin antibodies [21], and is typical for experimentally-induced secondary response antibodies [22].

Interpreting the breadth of spectrotype as evidence of somatic mutation, however, must be treated with caution. If the test antigen is complex, as is certainly the case for Sm, or has multiple epitopes, as is the case for most other autoantigens (see below), the broad spectrotype could simply reflect the sum of several unmutated lines directed against different proteins of a particle or different epitopes of a single protein. Against this improbable interpretation is that, at least in the case of anti-Jo-1 antibodies, the spectrotype pattern is unique in each patient even though all sera seem to react with the same epitopes [16].

The other evidence in favour of the view that somatic mutation has played a role in the development of autoantibodies is even less direct. For anti-La antibodies at least, the idiotypes of antibodies from individual patients are private [23]. This, of course, is consistent with somatic mutation but also with the employment of different germline V regions by different patients. Similarly, the absence of germline idiotypes characteristic of monoclonal lymphoma-derived rheumatoid factors [24] might mean no more than that those V regions are not the ones used in the production of disease-related rheumatoid factors.

The reason for invoking T cell help is also indirect. It mainly rests upon the predominance of IgG1 antibodies in some autoantisera and upon the above-mentioned evidence for isotype and class-switching. Until a more complete picture of the mechanisms for generating class-switching and somatic mutation exists, the role of T cells in driving a particular response cannot be accurately assessed.

In sum, then, the isotype, spectrotype, and idiotype profiles of the autoantibodies from individual patients with autoimmune diseases suggest that they are the antibodies of a secondary, antigen-driven response, and this supposition is strengthened by the differences found from patient to patient. In the more closely examined case of autoimmune mice, there is direct evidence for somatic mutation and, as noted below, for affinity maturation.

The Nature of the Antigen

The real starting point in the analysis of autoantibodies is the identified target antigen because it is the fact that an antiserum reacts with a self component that makes it an autoantiserum (Table 9.1). In the majority of the examples in autoimmune disease the antigenic target is intracellular and therefore unlikely to be exposed in a way that allows it to be the object of pathogenetic attack, and consequently it is not at all an empty question to ask if the reactivity observed is any more than a coincidence. In this section I will present the evidence to support the view that the identified targets of disease-associated autoantibodies are also the antigens that sustain the immune response.

Table 9.1. The role of antigen in an immune response

Antigen	Role	In autoimmunity
Initiating[a]	Starts B cell differentiation and proliferation	Unknown
Selecting[a]	Selects B cells making better fitting antibodies and prepares for memory response	Unknown
Sustaining[a]	Allows continued synthesis of antibodies if present, possibly spreading to other epitopes	Autoantigen
Target[b]	What the antibodies react with	Autoantigen

[a] Traditionally, the "immunogen".
[b] Traditionally, the "antigen".

In an immune response initiated by a deliberately injected antigen, the antigen that originally drives B cell division, drives class-switching, turns on the somatic mutation mechanism, turns off the somatic mutation mechanism, drives the cell division of successful clones, calls up those cells again in a secondary response, and reacts most strongly and with a rising affinity is one and the same. In analysing the end product of a fully developed response, however, one begins by knowing *only* a target antigen.

It is legitimate, therefore, to ask whether the identified target antigen is also the initiating, selecting, and sustaining antigens. In this section I will draw largely upon very recent studies with cloned autoantigens which establish the following generalizations about the identified target autoantigens and the autoantibodies directed against them:

1. Autoantibodies inhibit the function of their target antigens
2. They usually interact with more than one part (epitope) of a single protein
3. They usually interact with more than one protein if the target is a multiprotein particle
4. They usually react with different epitopes from those which a deliberately immunized animal will form antibodies against
5. Native may be preferred to denatured target
6. The autologous protein is usually preferred over the homologous protein from other species although the reactive epitopes may be present in a variety of species
7. The target epitopes are rarely short peptides but are large enough so that secondary structure is not only present but is also critical to the assembly of the epitope, and furthermore that some epitopes require the juxtaposition of distant parts of the same molecule

Together, this evidence powerfully implicates the target autoantigen in sustaining the autoimmune response but does not establish that the antigen initiated the response or drove the critical early differentiative steps.

Inhibition of function has been clearly demonstrated for some of the antisera with antibodies directed against the splicing mechanism, (anti-Sm and anti-RNP) [25]; for all of the antibodies against amino-acyl-tRNA synthetases [26–29]; for at least one other myositis-related antibody directed against another translational component [30]; for antibodies directed against cyclin, also known as the proliferating cell nuclear antigen (PCNA) [31]; for antibodies in scleroderma patients (Scl-70) directed against topoisomerase I [32]; for antibodies directed at poly (ADP-ribose) polymerase [33]; and for several of the antibodies of the antimitochondrial family directed against recognized enzymes [34,35]. For some target antigens whose precise function has not yet been determined, such as Ro and La, it has not been possible to test inhibition. In almost all the cases cited, the structural target was what was first recognized (as a positive ANA, for example, or a positive anti-mitochondrial antiserum by immunofluorescence) so the inhibition of function is not a trivial consequence of detection. Furthermore, in a few cases, animals immunized with the same antigens make antibodies which do not inhibit the function of the target [29] (F. W. Miller, 1989, unpublished work).

The identification of epitopes on a protein cannot be an exact science since the definition of a positive antigen–antibody reaction depends entirely upon the test system, the range of conformations of the test antigen offered to the antibodies, the sensitivity of detection, and a myriad of other factors. For example, most of the studies rely upon fusion proteins, upon synthesized peptides, upon fragmented proteins, and upon solid phase detection readouts. The relationship of the conformation of the test epitope to the native conformation is essentially unknowable at present. Nevertheless, from the available studies one may prudently infer that multiple epitopes are present on La [36–39], on Ro [40], on poly(ADP-ribose) polymerase [33], on Ku [41], and on several components of the splicing complex – A [42], A′ [43] B/B′ [44], B″ [45], and 68/70 kD [46]. There is, furthermore, evidence that more than one component of a target particle is recognized by antisera against the splicing particles, against ribosomes [47], against the signal recognition particle [48], against histones [49], against the centromere [50], and, in a couple of special cases, against both alanyl tRNA and alanyl-tRNA synthetase [28], and against both U1 RNA and the U1 binding protein, A [51,52]. In the latter two cases, the antibody pairs may be related as an idiotype–anti-idiotype pair [53]. This spreading – from epitope to epitope and from protein to protein – seems to me incontestable evidence for a sustaining role for the identified target.

In only a few instances has it been attempted to immunize an animal which will naturally develop an autoantibody in the course of autoimmune disease with the same antigen before the disease has developed. Nevertheless, when other species are immunized with the targets of human disease-related autoantibodies, they produce antibodies which, on the whole, differ from the human antibodies. Most strikingly, the antibodies may not inhibit the function of the target antigen [29]. This observation highlights the importance of the observed functional inhibition of the human autoantibodies. If the human with autoimmune disease were being immunized by, for example, the bacterial homologue of the human target antigen, the functional site is most likely to be conserved and consequently least likely to be perceived by the immune system as foreign. A human immunized with anything but the human protein or particle would be likely, just like the mouse immunized with the heterologous antigen, to make antibodies against everything but the active site. Why the active site is so often selected, then, needs to be understood.

The evidence on the range of homologous proteins from other species with which a human autoantibody will react is limited and not uniform. In the case of some targets, e.g. ribosomes, La [33,54], the range of species recognized is vast; in the case of some, e.g. Jo-1 [55], reactivity drops off as evolutionary distance increases; in the interesting case of Ku, not even other mammalian targets are recognized [56]; and most remarkably, in the case of the human antibodies which recognize the synaptonemal complex, the reaction with other species exceeds that with the human antigen [57], and, in one example, the serum recognized the newt complex better than the mouse complex, and the human complex not at all.

The actual delineation of epitopes, as discussed above, is not an exact science. Nevertheless, some interesting observations have been made. In at least three instances an attempt to locate a linear epitope in a target autoantigen by the synthesis of overlapping sequences by the pepscan method have failed to turn up any "linear" epitopes. The 15-mer peptides of PCNA overlapping by five amino acids did not react with any anti-PCNA sera [58]; all 225 possible nonamer peptides of B″ did not react with sera against U2 snRNPs (W. J. Habets, 1989, personal communication); and all 506 possible hexapeptides of histidyl-tRNA synthetase did not react with any anti-Jo-1 antisera [55]. By contrast, the experimental autoimmune uveitis induced by immunization with a heterologous retinal protein, the pepscan method successfully identified an apparently critical linear epitope on the immunizing protein [59]. When an apparent epitope has been identified by peptide synthesis, it is usually substantially larger. The smallest peptide epitopes so far identified are a 22-mer in ribosomal P proteins [60], and a 20-mer in the mitochondrial dihydrolipoamide acyltransferase [34]. Many other larger reactive peptides, usually made as fusion proteins, have been identified, but it is not possible to distinguish the region which contacts the antigen-binding site of the antibody from those regions necessary to hold the contact amino acids in the correct conformation.

In summary, then, the evidence strongly suggests that the native target antigen, whether it is a single protein or a complex particle, plays a critical role in sustaining the autoantibody response.

From what is known of the surface of protein molecules it is hardly surprising to discover that an autoantibody which reacts with an intact, functional protein should require a particular conformation [61,62]. It is, however, instructive to reflect upon the implications of these observations. The B cell must be seeing the native, functional protein. If not, it must be seeing some structure whose three-dimensional conformation is very like the conformation of the target antigen. A short peptide is highly unlikely to present the same surface to the immune system if it occurs in two proteins with otherwise unrelated primary structures. Conversely, as in the example of the cross-reaction between B/B′ and D of the Sm particle, there can be cross-reaction, even by a hybridoma, between proteins without recognizable homology at the level of primary structure. It is now apparent, for example, that carboyhydrates and peptides may present immunologically equivalent surfaces [63].

I believe, therefore, that the kind of search for molecular mimicry which is possible in the memory banks of computers is unlikely to reveal structures from the proteins of other species responsible for generating autoantibodies. That does not at all mean that I disbelieve in molecular mimicry or some variant of it such as that I have previously suggested [53]. It does mean, however, that the search will be more difficult and will require deeper insights into molecular structures and interactions. The search must be for shared conformation, as evidenced by shared function, shared binding ability, or shared idiotype.

Finally, as the immunological response represented by the autoantiserum is traced back to its origins, it is necessary to consider whether the identified target antigen qualifies as the original stimulating antigen. It is here that the evidence begins to fail us, and in fact, there is very little evidence at all from which to judge. The only direct evidence that a particular antigen has been responsible for driving a particular immune response would be that there is, over the course of the response, affinity maturation. That is, that the antibodies late in the response are better able to react with the antigen than those early in the response. Only in the case of mouse anti-DNA antibodies has this been clearly demonstrated through the use of hybridomas, sequencing, and careful measures of affinity [4]. Since affinity maturation is a very early event in an immune response, the prospects are not bright for being able to observe it in most patients with autoimmune diseases whose autoantibodies are present and well-developed by the time they reach the doctor's office.

The situation, however, is not hopeless. For example, ter Borg and colleagues have watched the unfolding of an anti-Sm response in a patient with systemic lupus erythematosus and shown beautifully that the response began with a reaction to only one protein and then spread to other proteins of the same particle [64]. Identifying the first epitope recognized on that protein will be highly interesting. There is also evidence of epitope progression in anti-La antibodies [65]. In neither of these cases, however, has there yet been an attempt to measure changes in affinity over time. This kind of observation should, when possible, be repeated in other patients. Naturally, understanding the conditions which allow spreading to occur and which signal a B cell to secrete an autoantibody will be revealing, too.

The fortuitous presentation to our clinic of a patient with anti-Jo-1-positive myositis on whom extensive early sera are available will allow us to study the development of this autoantibody from this perspective. Careful studies of the early stages of anti-ribosomal P proteins and anti-Sm in MRL mice will also be critical to determining whether the target antigen is indeed the antigen driving B cell selection in the early stages of the response.

Finally, almost nothing is known about the response of T cells to the target autoantigens. Their role, if any, in permitting or encouraging B cells to secrete autoantibodies awaits exploration.

References

1. Nossal GJ (1989) Immunologic tolerance: collaboration between antigen and lymphokines. Science 345:147–153
2. Goodnow CC, Crosbie J, Adelstein S et al. (1988) Altered immunoglobulin expression and functional silencing of self-reactive B lymphocytes in transgenic mice. Nature 334:676–682
3. Nemazee DA, Büurki K (1989) Clonal deletion of B lymphocytes in a transgenic mouse bearing anti-MHC class I antibody genes. Nature 337:562–566
4. Shlomchik MJ, Aucoin AH, Pisetsky DS, Weigert MG (1987) Structure and function of anti-DNA autoantibodies derived from a single autoimmune mouse. Proc Natl Acad Sci USA 84:9150–9154
5. Marion TN, Bothwell AL, Briles DE, Janeway CA Jr (1989) IgG anti-DNA autoantibodies within an individual autoimmune mouse are the products of clonal selection. J Immunol 142:4269–4274
6. Tojo T, Friou GJ, Spiegelberg HL (1979) Immunoglobulin G subclass of human antinuclear antibodies. Clin Exp Immunol 6:145–151
7. Lieu TS, Reimer CB, Sontheimer RD (1988) Immunoglobulin class and subclass profile of the Ro/SS-A autoantibody response. J Invest Dermatol 90:158–164
8. Pearce DC, Yount WJ, Eisenberg RA (1986) Subclass restriction of anti-SS-B (La) autoantibodies. Clin Immunol Immumopathol 38:111–119
9. Davies TF, Weber CM, Wallack P, Platzer M (1986) Restricted heterogeneity and T cell dependence of human thyroid autoantibody immunoglobulin G subclasses. J Clin Endocrinol Metab 62:945–949
10. McLachlan SM, Feldt-Rasmussen U, Young ET et al. (1987) IgG subclass distribution of thyroid autoantibodies: 'fingerprint' of an individual's response to thyroglobulin and thyroid microsomal antigen. Clin Endocrinol (Oxf) 26:335–346
11. Bonfa E, Chu JL, Brot N, Elkon KB (1987) Lupus anti-ribosomal P peptide antibodies show limited heterogeneity and are predominantly of the IgG1 and IgG2 subclasses. Clin Immunol Immunopathol 45:129–138
12. Jones CC, Hamilton RG, Jordon RE (1988) Subclass distribution of human IgG autoantibodies in pemphigus. J Clin Immunol 8:43–49
13. Eisenberg RA, Dyer K, Craven SY, Fuller CR, Yount WJ (1985) Subclass restriction and

polyclonality of the systemic lupus erythematosus marker antibody anti-Sm. J Clin Invest 75:1270–1277
14. Rødgaard A, Nielsen FC, Djurup R, Somnier F, Gammeltoft S (1987) Acetylcholine receptor antibody in myasthenia gravis: predominance of IgG subclasses 1 and 3. Clin Exp Immunol 67:82–88
15. Cohen PL, Cheek RL, Hadler JA, Yount WJ, Eisenberg RA (1987) The subclass distribution of human IgG rheumatoid factor. J Immunol 139:1466–1471
16. Miller RW, Twitty SA, Biswas T, Plotz PH (1990) The origin and regulation of a disease-specific autoantibody response: antigenic epitopes, spectrotype stability, and isotype restriction of anti-Jo-1 autoantibodies. J Clin Invest 85:468–475
17. Siekevitz M, Kocks C, Rajewsky K, Dildrop R (1987) Analysis of somatic mutation and class switching in naive and memory B cells generating adoptive primary and secondary responses. Cell 48:757–770
18. Shlomchik MJ, Marshak-Rothstein A, Wolfowicz CB, Rothstein LT, Weigert MG (1987) The role of clonal selection and somatic mutation in autoimmunity. Nature 328:805–811
19. Behar SM, Scharff MD (1988) Somatic diversification of the S107 (T15) VH11 germ-line gene that encodes the heavy-chain variable region of antibodies to double-stranded DNA in (NZB × NZW) F1 mice. Proc Natl Acad Sci USA 85:3970–3974
20. Eisenberg RA, Craven SY, Cohen PL (1987) Isotype progression and clonality of anti-Sm autoantibodies in MRL/Mp-*lpr*/*lpr* mice. J Immunol 139:728–733
21. Delves PJ, Roitt IM (1988) Long term spectrotypic and idiotypic stability of thyroglobulin autoantibodies in patients with Hashimoto's thryoiditis. Clin Exp Immunol 71:459–463
22. Askonas BA, Williamson AR, Wright BEG (1970) Selection of a single antibody-forming cell clone and its propagation in syngeneic mice. Proc Natl Acad Sci USA 67:1398–1403
23. Horsfall AC, Venables PJW, Mumford PA, Maini RN (1986) Idiotypes on antibodies to the La(SS-B) antigen are restricted and associated with the antigen binding site. Clin Exp Immunol 63:395–401
24. Fong S, Chen P, Gilbertson T, Weber J, Fox R, Carson D (1986) Expression of three cross-reactive idiotypes on rheumatoid factor from patients with autoimmune diseases and seropositive adults. J Immunol 137:122–128
25. Bozzoni I, Annesi F, Beccari E, Fragapane P, Pierandrei-Amaldi P, Amaldi F (1984) Splicing of Xenopus laevis ribosomal protein RNAs is inhibited in vivo by antisera to ribonucleoproteins containing U1 small nuclear RNA. J Molec Biol 180:1173–1178
26. Mathews MB, Bernstein RM (1983) Myositis autoantibody inhibits histidyl-rRNA synthetase: a model for autoimmunity. Nature 304:177–179
27. Mathews MB, Reichlin M, Hughes GR, Bernstein RM (1984) Anti-threonyl-tRNA synthetase, a second myositis-related autoantibody. J Exp Med 160:420–434
28. Bunn CC, Bernstein RM, Mathews MB (1986) Autoantibodies against alanyl-tRNA synthetase and tRNA(ALA) coexist and are associated with myositis. J Exp Med 163:1281–1291
29. Dang CV, Tan EM, Traugh JA (1988) Myositis autoantibody reactivity and catalytic function of threonyl-tRNA synthetase. FASEB J 2:2376–2379
30. Targoff IN, Arnett FC, Berman L, O'Brien C, Reichlin M (1989) Anti-KJ: a new antibody associated with the syndrome of polymyositis and interstitial anti-centromere antibodies. J Clin Invest 77:426–430
31. Prelich G, Tan CK, Kostura M et al. (1987) Functional identity of proliferating cell nuclear antigen and a DNA polymerase-delta auxiliary protein. Nature 326:517–520
32. Shero JH, Bordwell B, Rothfield NF, Earnshaw WC (1986) High titers of autoantibodies to topoisomerase I (Scl-70) in sera from scleroderma patients. Science 231:737–740
33. Yamanaka H, Willis EH, Carson DA (1989) Human autoantibodies to poly(adenosine diphosphate-ribose) polymerase recognize cross-reactive epitopes associated with the catalytic site of the enzyme. J Clin Invest 83:180–186
34. Van de Water J, Gershwin ME, Leung P, Ansari A, Coppel RL (1988) The autoepitope of the 74-kD mitochondrial autoantigen of primary biliary cirrhosis corresponds to the functional site of dihydrolipoamide acetyltransferase. J Exp Med 167:1791–1799
35. Van de Water J, Fregeau D, Davis P et al. (1988) Autoantibodies of primary biliary cirrhosis recognize dihydrolipamide acetyltransferase and inhibit enzyme function. J Immunol 141:2321–2324
36. Chan EK, Tan EM (1986) Epitopes and structural domains of a RNA-binding nuclear protein, SS-B/La: similarities with adenovirus DNA-binding protein. Scand J Rheumatol (Suppl) 61:102–105

37. Chan EK, Francoeur AM, Tan EM (1986) Epitopes, structural domains, and asymmetry of amino acid residues in SS-B/La nuclear protein. J Immunol 136:3744–3749
38. St Clair EW, Pisetsky DS, Reich CF, Keene JD (1988) Analysis of autoantibody binding to different regions of the human La antigen expressed in recombinant fusion proteins. J Immunol 141:4173–4180
39. Rauh AJ, Hornig H, Lüuhrmann R (1988) At least three distinct B cell epitopes reside in the C-terminal half of La protein, as determined by a recombinant DNA approach. Eur J Immunol 18:2049–2057
40. Lieu T-S, McCauliffe DP, Arnett FC et al. (1989) Epitope mapping of the human Wil-2 cell Ro/SS-A (Ro) autoantigenic polypeptide. Arthritis Rheum 32:S26 (abstract)
41. Reeves WH, Sthoeger ZM (1989) Molecular cloning of cDNA encoding the p70 (Ku) lupus autoantigen. J Biol Chem 264:5047–5052
42. Habets WJ, Sillekens PT, Hoet MH, McAllister G, Lerner MR, van Venrooij WJ (1989) Small nuclear RNA-associated proteins are immunologically related as revealed by mapping of autoimmune reactive B-cell epitopes. Proc Natl Acad Sci USA 86:4674–4678
43. Fresco LD, Keene JD (1989) Leucine periodicity of U2 snRNP autoantigen A′ implicated in U2 snRNP protein:protein interactions. Sitzungsberichte Heidelberger Akad Wissenschaften Mathematisch-naturwissenschaftliche Klasse 4:15–16 (abstract)
44. Takeda Y, Wang GS, Sharp GC (1989) Immunological characterization of IgM class autoantibodies against Sm-antigenic B′/B and D polypeptides of U small nuclear riboproteins (snRNPs). Sitzungsberichte Heidelberger Akad Wissenschaften. Mathematisch-naturwissenschaftliche Klasse 4:92–94 (abstract)
45. Habets WJ (1989) Mapping of autoimmune reactive sites on snRNP proteins. Sitzungsberichte Heidelberger Akad Wissenschaften Mathematisch-naturwissenschaftliche Klasse 4:26–27 (abstract)
46. Guldner HH, Netter HJ, Szostecki C, Lakomek HJ, Will H (1988) Epitope mapping with a recombinant human 68-kDa (U1) ribonucleoprotein antigen reveals heterogeneous autoantibody profiles in human autoimmune sera. J Immunol 141:469–475
47. Bonfa E, Parnassa AP, Rhoads DD, Roufa DJ, Wool IG, Elkon KB (1989) Anti-ribosome S10 antibodies in patients with lupus and MRL/*lpr* mice. Arthritis Rheum 32:S20 (abstract)
48. Targoff IN, Johnson AE, Miller FW (1988) Polymyositis and antibody to the signal recognition particle. Clin Res 36:537A (abstract)
49. Rubin RL, Burlingame RW (1989) Drug induced anti-histone antibodies reflect nucleosomal substructures. Sitzungsberichte Heidelberger Akad Wissenschaften Mathematisch-naturwissenschaftliche Klasse 4:76–78 (abstract)
50. Earnshaw W, Bordwell B, Marino C, Rothfield N (1986) Three human chromosomal autoantigens are recognized by sera from patients with lung disease. J Clin Invest 84:162–172
51. Deutscher SL, Keene JD (1988) A sequence-specific conformational epitope on U1 RNA is recognized by a unique autoantibody. Proc Natl Acad Sci USA 85:3299–3303
52. Lutz-Freyermuth C, Keene JD (1989) The U1 RNA-binding site of the U1 small nuclear ribonucleoprotein (snRNP)-associated A protein suggests a similarity with U2 snRNPs. Molec Cell Biol 9:2975–2982
53. Plotz PH (1983) Autoantibodies are anti-idiotype antibodies to antiviral antibodies. Lancet ii:824–826
54. Chan EK, Tan EM (1987) Human autoantibody-reactive epitopes of SS-B/La are highly conserved in comparison with epitopes recognized by murine monoclonal antibodies. J Exp Med 166:1627–1640
55. Waite K, Miller FW, Plotz PH (1989) Anti-Jo-1 antibodies are directed at an evolutionarily-conserved conformational site on human histidyl-tRNA synthetase. Sitzungsberichte Heidelberger Akad Wissenschaften Mathematisch-naturwissenschaftliche Klasse 4:100–101 (abstract)
56. Francoeur AM, Peebles CL, Gompper PT, Tan EM (1986) Identification of Ki (Ku, p70/p80) autoantigens and analysis of anti-Ki autoantibody reactivity. J Immunol 136:1648–1653
57. Haaf T, Machens A, Schmid M (1989) Human autoantibodies to synaptonemal complex. Sitzungsberichte Heidelberger Akad Wissenschaften Mathematisch-naturwissenschaftliche Klasse 4:24–25 (abstract)
58. Huff JP, Roos G, Peebles CL et al. (1989) Structural analysis of the epitopic sites of PCNA/cyclin. Arthritis Rheum 32:S80 (abstract)
59. Stiemer R, Gausepohl H, Mirshahi M et al. (1989) Localization of an immunopathologically important epitope in the bovine retinal S-antigen by the pepscan method. Sitzungsberichte Heidelberger Akad Wissenschaften Mathematisch-naturwissenschaftliche Klasse 4:88–91 (abstract)

60. Elkon K, Skelly S, Parnassa A et al. (1986) Identification and chemical synthesis of a ribosomal protein antigenic determinant in systemic lupus erythematosus. Proc Natl Acad Sci USA 83:7419–7423
61. Berzofsky JA (1985) Intrinsic and extrinsic factors in protein antigenic structure. Science 229:932–940
62. Barlow DJ, Edwards MS, Thornton JM (1986) Continuous and discontinuous protein antigenic determinants. Nature 322:747–748
63. Diakun KR, Matta KL (1989) Synthetic antigens as immunogens: Part III. Specificity analysis of an anti-anti-idiotypic antibody to a carbohydrate tumor-associated antigen. J Immunol 142:2037–2040
64. ter Borg EJ, Horst G, Hummel E, Jaarsma D, Limburg PC, Kallenberg CG (1988) Sequential development of antibodies to specific Sm polypeptides in a patient with systemic lupus erythematosus: evidence for independent regulation of anti-double-stranded DNA and anti-Sm antibody production. Arthritis Rheum 31:1563–1567
65. Yamamoto K, Kohsaka H, Miura H, et al. (1989) Epitope progression in autoantibody production suggested by epitope mappings of SS-B/La and U2snRNP B″ Protein. Sitzungsberichte Heidelberger Akad Wissenschaften Mathematisch-naturwissenschaftliche Klasse 4:107–108 (abstract)

Chapter 10

What are the Immunogens That Incite the Autoantibodies of Systemic Lupus Erythematosus?

T. Koizumi, A. Puccetti, J. André-Schwartz, K. Barrett and R. S. Schwartz

Anti-DNA Antibodies

A major problem in the field of autoimmunity is the identification of immunogens reponsible for the production of pathogenic autoantibodies. Lack of knowledge of disease-provoking immunogens is a serious impediment to the development of specific means of preventing and treating autoimmune disorders. In systemic lupus erythematosus, rheumatoid arthritis, type 1 diabetes and multiple sclerosis, some of the principal autoimmune diseases, the aetiological immunogen is unknown. Advances in the molecular characterization of autoantigens, mainly by using autoantibodies to identify cDNA clones of autoantigens (reviewed in [1]) have been substantial, but whether those cloned antigens can instigate pathogenic autoimmune mechanisms is unknown.

Systemic lupus erythematosus is an important example of the gap in our knowledge of disease-causing immunogens. The emblematic autoantibody of lupus binds to double strand DNA, but this nucleic acid antigen is barely immunogenic even in MRL-*lpr*/*lpr* mice, which are genetically programmed to produce spontaneously anti-DNA antibodies [2]. The failure of these lupus-prone mice, and of normal strains as well, to respond immunologically to native DNA is not due to a lack of immunogenicity of nucleic acids generally. Other forms of DNA, such as Z-DNA (a double strand molecule with a leftward spiralling helix instead of the usual right-handed helical structure) are highly immunogenic in both normal and MRL-*lpr*/*lpr* mice [2,3]. Moreover, injection of double strand DNA has not been shown to cause any disease in experimental animals. This is why there is no animal model of induced systemic lupus erythematosus. An animal model of the disease had to await the discovery of spontaneous systemic lupus erythematosus in NZB mice [4].

It may be that DNA must be presented to the immune system in the form of a complex with a particular kind of protein, as would occur in cell nuclei, in order to promote its immunogenicity. One reason why "naked" double strand DNA is not immunogeneic is that it is rapidly degraded by serum nucleases [5]. A "foreign" DNA might also provide the immunogenic stimulus in systemic lupus erythematosus. The latter has been suggested by the work of Gilkeson et al. [6], who found antibodies against double strand DNA in mice immunized with double strand DNA from *E. coli*, but not from calf thymus. However, those antibodies were unlike the anti-DNA antibodies in systemic lupus erythematosus in that they were highly specific for the bacterial DNA; lupus anti-DNA antibodies react with DNA from all species [7].

The origin of anti-DNA antibodies has also been attributed to polyclonal B cell activation, an antigen-independent mechanism in which B cells are non-specifically activated [8]. The high frequency of B cells in the normal immune repertoire with the capacity to produce anti-DNA antibodies [9] and the highly abnormal, activated T cell population in MRL-*lpr/lpr* mice [10] – which would provide the lymphokines required for polyclonal B cell activation [11] – would be consistent with that view. However, molecular studies of anti-DNA antibodies from MRL-*lpr/lpr* mice have shown that they have the features expected of an antigen-driven immune response: they are derived from only a few clones and their V genes are mutated [12].

We are thus left with a puzzle. The molecular evidence supporting the origin of anti-DNA antibodies via a conventional immune response (selection of clones by an immunogen) is persuasive, at least in MRL-*lpr/lpr* mice, yet the immunogen that might be responsible for the findings defies identification. We therefore need to consider the possibility that anti-DNA antibodies might arise by a mechanism that is entirely independent of immunization with DNA or, for that matter, with any other nucleic acid. At first glance, that idea may seem curious, but it does reflect a fact generally known in immunology: the immunogen and the antigen are not always the same. In the Weil–Felix reaction, to take an historic example, agglutinins against the OX19 strain of *Proteus vulgaris*, a bacillus, occur in the serum of patients with rickettsial diseases. In principle, therefore, a molecule other than DNA could have the potential to incide anti-DNA antibodies in patients with systemic lupus erythematosus. Immunochemical examination of monoclonal anti-DNA antibodies derived from MRL-*lpr/lpr* mice or from humans with systemic lupus erythematosus has revealed a high degree of cross-reactions, often with unexpected ligands [3,7], and in some cases a synthetic polynucleotide (especially poly (dT)) is preferred over DNA [13]. In at least one well-studied case, cardiolipin, which is known to cross-react with DNA because of the array of phosphate groups in its backbone [14], has been shown to induce the production of anti-DNA antibodies when injected into normal animals [15].

Autoantibodies Against Ribonuclearproteins

Before proceeding, we need to consider another system of autoantibodies in systemic lupus erythematosus, the autoantibodies against ribonucleoproteins.

The corresponding autoantigens are highly relevant to some of our current work with anti-DNA antibodies. There are two categories of autoantibodies against these particles: anti-RNP antibodies, which bind the U1 snRNP-specific polypeptides 70K, A and C, and anti-Sm antibodies, which bind to a shared epitope in the B′, B″ and D polypeptides. U1 RNP is a member of a group of uridine-rich small nuclear ribonuceloprotein (snRNP) particles that participate in pre-mRNA splicing [16]. The U1 RNP particle consists of the U1 small RNA and at least nine associated polypeptides [17,18]. Six of these polypeptides are common to the U1, U2, U4, U5 and U6 snRNPs; they are termed B′ (Mr = 29 kDa), B (Mr = 28 kDa), D (Mr = 16 kDa), and E-G (Mr = 13-11 kDa). Three of the nine polypeptides are unique to the U1 RNP particle: 70K (Mr = 70 kDa, also referred to as 68K), A (Mr = 33 kDa), and C (Mr = 22 kDa). The U2 RNP particle contains two unique polypeptides, A′ and B′ (26 500 kDa) [19].

Several components of snRNP autoantigens have been cloned and sequenced, and corresponding recombinant proteins have been produced. These include the 70K polypeptide [20–22], the E antigen of the Sm complex [23], and Sm-D antigen [24], the C polypeptide of the U1 RNA particle [25], and the B′ polypeptide shared by U1 and U2 particles [26]. The major epitopes of the 70K autoantigen are being elucidated with serological tests of the recombinant antigen and fusion polypeptides. A region of approximately 30 amino acids in the amino-terminal part of the molecule defines reactivity with 16/17 human anti-70K antisera. This region of the molecule is especially important in light of our preliminary results, discussed below.

There is a serological relationship between DNA and RNP, however. Monoclonal antibodies that bind to both autoantigens have been described [27] and, relevant to our work, anti-DNA and anti-RNP antibodies sharing the same idiotype have been found both in mice [28,29] and human serum [30]. The role, if any, of anti-Sm and anti-RNP autoantibodies in the pathogenesis of autoimmune diseases is unknown. That anti-RNP antibodies can be pathogenic is suggested by the report of two infants with the neonatal lupus syndrome who, along with their mothers, had anti-RNP, but not anti-Sm, antibodies in their sera. The mothers had systemic lupus; both infants had an extensive rash which waned as the transplacentally transmitted anti-RNP antibodies were catabolized [31]. The development of severe dermatitis in these infants is of interest because U1 RNP antigens appear on the plasma membrane of epidemeral keratinocytes after exposure to ultraviolet light in vitro [32].

The origins of anti-Sm and anti-RNP autoantibodies are unknown. The involvement of T cells in the production of anti-Sm antibodies can be surmised from the subclass restriction of the autoantibodies in humans (IgG1) [33] and in MRL-*lpr*/*lpr* mice (IgG2a) [34]. Release of Sm and RNP antigens from the nuclei of dead or injured cells may, according to one hypothesis, render them immunogenic [35]. The finding that normal mice respond to immunization with purified Sm [36] and U1 RNP polypeptides [37] by producing anti-Sm/RNP antibodies of the type found in systemic lupus supports the hypothesis that immunization by endogenous nuclear autoantigens causes production of the autoantibodies. However, this idea does not explain why the usual daily turnover of billions of nucleated cells does elicit anti-Sm and anti-RNP in normal subjects.

Recently, a cDNA-derived amino acid sequence of the 70K U1 RNP polypeptide was found to contain 23 residues homologous to a region of the $p30^{gag}$ protein of murine leukaemia viruses. The homology resides in an antigenic portion of the 70K polypeptide and occurs in a tandem repeat in the $p30^{gag}$ of most mammalian type C retroviruses. Furthermore, an antiserum against murine leukaemia virus $p30^{gag}$ protein reacted with anti-U1 snRNP antigens [38]. These findings raise the possibility of a retroviral origin of such autoantibodies.

Weening et al. [39] have induced immune complex nephritis and anti-RNP antibodies in mice with sublethal doses of mercuric chloride. The relevant nuclear antigen in this model has been found to be a U3 snRNP polypeptide – a characteristic autoantigen in scleroderma [40]. The mechanism by which mercuric chloride causes a major histocompatibility complex (MHC)-restricted production of autoantibodies against this nuclear antigen is unknown. One obvious possibility is that modification of the ribonucleoprotein by the heavy metal renders it immunogenic. However, mercuric chloride is known to cause polyclonal activation of both helper T cells and B cells [41], effects which may be of considerable importance.

A model of autoimmunization we originated, chronic graft-versus-host disease (GvHD) [42], has been investigated by Gelpi et al. [43]. They found that (BALB/c × A/J)F1 mice injected with A/J spleen cells developed nephritis, dermatitis and autoantibodies to components of the U1 and U3 antigens. The particulars of the autoimmune disease in this model are genetically controlled by both the donor and the host. Identical experimental protocols can lead to autoimmune haemolytic anaemia, immune-complex nephritis, a scleroderma-like disease, or malignant lymphoma in different donor–host combinations [44]. Polyclonal activation of T cells and B cells is prominent in the chronic GvHD; no exogenous immunogen is needed for the production of a variety of autoantibodies, nor is destruction of host cells by donor cells required for their elicitation [45].

Autoantigen-Limiting Idiotypes Stimulate the Production of Autoantibodies

Polyclonal B cell activation is also prominent in lupus-prone mice, including the MRL-*lpr*/*lpr* strain [44], but whether pathogenic autoantibodies arise from that mechanism is unclear. Non-specific expansion of the B cell compartment does not satisfactorily explain why a given autoimmune disease is dominated by distinctive autoantibody populations. Of two patients with systemic lupus, why does one develop nephritis and the other immune thrombocytopenia if polyclonal B cell activation is the thematic disturbance in both cases?

We have been investigating some of these problems with a remarkable system in which we have found evidence of structural and immunological relationships between an immunoglobulin idiotype and the U1 RNP nuclear autoantigen. This idiotype appears to be a component of a network in MRL-*lpr*/*lpr* mice in which antibody-1 (Ab-1) binds to the 70K U1 RNP polypeptide, and antibody-2 (Ab-2) mimics the U1 RNP autoantigen. Our data suggest that these two kinds of antibodies are frequent in MRL-*lpr*/*lpr* mice. A notable feature of the experimental model we are studying is that autoimmunization can be elicited in normal mice by injection of a purified immunoglobulin light chain dimer (light chain G). This light chain, and most likely others in our collection, has two distinctive properties: it defines an idiotypic marker recognized by Ab-1, and the amino acid sequence of its CDR3 is homologous to a sequence in an antigenic region of the 70K U1 RNP polypeptide. Our evidence shows that immunization of normal (BALB/c) mice with the purified MRL-*lpr*/*lpr* light chain G elicits the production of anti-RNP antibodies. Even more interesting is that those BALB/c mice also produce antibodies against double-strand DNA and develop immune deposit nephritis. This system has the potential of elucidating how pathogenic autoantibodies could arise by autoimmunization with idiotypes that mimic autoantigens. Studies of this model promise to define epitopes concerned with the production of pathogenic anti-DNA antibodies. The results we have already obtained raise the exciting prospect of new therapeutic approaches in systemic lupus.

Our results suggest the following chain of events: polyclonal B cell activation → expression of autoantigen-mimicking idiotypes → selection of autoantibody-producing clones by idiotype. The experimental model we are studying provides an unusual opportunity to test the hypothesis that the selective pressure giving rise to anti-DNA antibodies could be an idiotype [46]. Implicit in this hypothesis is the participation of somatically generated idiotypes, which can explain the perplexing, apparently random, events in many autoimmune diseases. The discordance of type 1 diabetes in identical twins raised in the same environment and the spontaneous remissions and exacerbations of systemic lupus [47] are but two examples. Another is that anti-Sm antibodies occur in only about 30% of patients with lupus. Anti-Sm antibodies develop in a similar minority of MRL-*lpr*/*lpr* mice even though they are an almost genetically uniform inbred strain. A thorough analysis has disclosed that stochastic events probably determine the production of anti-Sm autoantibodies in MRL-*lpr*/*lpr* mice; there was no evidence for genetic, environmental, gender or parental influences on the tendency of individual animals to produce those antibodies [48].

A topic of substantial current interest is whether idiotypic networks can influence the process of autoimmunization. Such networks may modulate, via auto-anti-idiotypes, the production of autoantibodies, or they might actually induce autoantibodies if the variable region of an Ab-2 imitates the structure or contact region of an autoantigen (the internal image effect [49]. An example of this phenomenon is the development of myasthenia gravis in rabbits immunized with BisQ, a synthetic compound that binds to the acetylcholine receptor. Ab-2 against anti-BisQ antibodies (Ab-1) in the serum of the affected rabbits had properties of anti-acetylcholine receptor antibodies [50]. Mice immunized with insulin can produce not only anti-insulin antibodies but also anti-(anti-insulin) antibodies which bind the insulin receptor [51]. Another example of the induction of autoimmunity via idiotypes was found in mice immunized with the human monoclonal anti-DNA antibody bearing the 16/6 idiotype. The animals produced high levels of anti-16/6 idiotype antibodies and also anti-(anti-16/6 idiotype) antibodies that had anti-DNA activity. Other autoantibodies, including anti-Sm/RNP, anti-Ro, anti-La and anti-cardiolipin autoantibodies were also found. Of even greater significance is the development of immune complex nephritis in these animals, with mouse 16/6 idiotype in the renal lesion [52]. In this model the human 16/6 idiotype (Ab-1) probably evoked a population of Ab-2s that mimicked the variable region of the human autoantibody; these in turn elicited a population of pathogenic Ab-3s with the 16/6 idiotype.

Our finding that the CDR3 of an Ab-2 is structurally homologous to an autoantigen has several precedents. Fougereau's group [53] analysed an idiotypic network evoked in mice by immunization with a random copolymer of glutamine, alanine and tyrosine (GAT): Ag → Ab-1 → Ab-2 → Ab-3. The Ab-1s and Ab-3s that expressed the dominant anti-GAT idiotype used closely related VH and Vκ genes. The D region of the Ab-2s contained either a Tyr-Tyr-Glu or a Glu-Glu-Tyr sequence, which are characteristic of the major epitopes of GAT. This D region may therefore contain an internal image of the immunogen. Bona et al. [54] have described an anti-idiotype that reacted with both an idiotype of a monoclonal IgG human rheumatoid factor and the Fc fragment of human IgG. They called this doubly reactive anti-idiotype an epibody – an antibody that recognizes both an idiotype and an epitope on the original antigen. Subsequently, Chen et al. [55] found a second example of an epibody: an antiserum against a synthetic peptide corresponding to the first hypervariable region of the κ light chain of a human rheumatoid factor was found to bind pooled normal human IgG. This cross reaction involving the VL region of an immunoglobulin and the Fc domain of IgG was attributed to a Ser-Ser-Ser peptide in both the γ chains of IgG and the light chain of the rheumatoid factor. Bruck et al. [56] reported that an Ab-2 against an anti-reovirus haemagglutinin antibody contained an amino acid sequence that was also present in the viral antigen.

References

1. Schwartz RS, Datta DK (1989) Autoimmunity and autoimmune diseases. In: Paul WE (ed) Fundamental immunology. Raven Press, New York, pp 819–866
2. Madaio MP, Hoder S, Stollar BD et al. (1984) Responsiveness of autoimmune and normal mice to nucleic acid antigens. J Immunol 132:872
3. Mackworth-Young C, Schwartz RS (1988) Autoantibodies to DNA. CRC Crit Rev Immunol 8:147
4. Bielschowsky M, Helyer BJ, Howie JB (1959) Spontaneous hemolytic anemia in mice of the NZB/BI strain. Proc Univ Otago Med Sch 37:9
5. Braun RP, Lee JS (1988) Immunogenic duplex nucleic acids are nuclease resistant. J Immunol 141:2084
6. Gilkeson GS, Grudier JP, Karounos DG et al. (1989) Induction of anti-double-stranded DNA antibodies in normal mice by immunization with bacterial DNA. J Immunol 142:1489
7. Shoenfeld Y, André-Schwartz J, Stollar BD et al. (1987) Anti-DNA antibodies. In: Lahita RG (ed) Systemic lupus erythematosus. John Wiley, New York, p 213
8. Klinman D, Steinberg AD (1987) Systemic autoimmune disease arises from polyclonal B-cell activation. J Exp Med 165:1755
9. Souroujon M, White-Scharff ME, André-Schwartz J et al. (1988) Preferential autoantibody reactivity of the preimmune B cell repertoire in normal mice. J Immunol 140:4173
10. Datta SK, Patel H, Berry D (1987) Induction of a cationic shift in IgG anti-DNA autoantibodies. Role of T helper cells with classical and novel phenotypes in three murine models of lupus nephritis. J Exp Med 165:1252
11. Dobashi K, Ono S, Mirakami S et al. (1987) Polyclonal B cell activation by a B cell differentiation factor, B151-TRF2 IV. B151-TRF2 as a B cell differentiation factor closely associated with autoimmune disease. J Immunol 138:780
12. Shlomchik MJ, Aucoin AH, Pisetsky DS et al. (1987) Structure and function of anti-DNA autoantibodies derived from a single autoimmune mouse. Proc Natl Acad Sci USA 84:9150
13. Schwartz RS, Stollar BD (1985) Origins of anti-DNA autoantibodies. J Clin Invest 75:321
14. Lafer EM, Rauch J, Andrzejewski C Jr et al. (1981) Polyspecific monoclonal lupus autoantibodies reactive with both polynucleotides and phospholipids. J Exp Med 153:879
15. Rauch J, Tannenbaum H, Stollar BD et al. (1984) Mononclonal anti-cardiolipin antibodies bind to DNA. Eur J Immunol 14:529

16. Dreyfuss G, Philipson L, Mattaj IW (1988) Ribonucleoprotein particles in cellular processes. J Cell Biol 106:1419
17. Lerner MR, Steitz JA (1979) Antibodies to small nuclear RNAs complexed with proteins are produced by patients with systemic lupus erythematosus. Proc Natl Acad Sci USA 76:5495
18. Pettersson I, Hinterberger M, Mimori T (1984) The structure of mammalian small nuclear ribonucleoproteins. J Biol Chem 259:12850
19. Tan EM (1989) Interactions between autoimmunity and molecular cell biology: Bridges between clinical and basic science. J Clin Invest 84:1
20. Theissen H, Etzerodt M, Reuter R (1986) Cloning of the human cDNA for the U1 RNA-associated 70K protein. EMBO J 5:3209
21. Spritz RA, Strunk K, Surowy CS (1987) The human U1-70K snRNP protein: cDNA cloning, chromosomal localization, expression, alternative splicing and RNA-binding. Nucl Acids Res 15:1:373
22. Netter JJ, Guldner HH, Szostecki C (1988) A recombinant autoantigen derived from the human U1 small nuclear RNP-specific 68-kd protein. Arthritis Rheum 31:616
23. Wieben ED, Rohleder AM, Nenninger JM (1985) cDNA cloning of a human autoimmune nuclear ribonucleoprotein antigen. Proc Natl Acad Sci USA 82:7914
24. Rokeach LA, Haselby JA, Hoch SO (1988) Molecular cloning of a cDNA encoding the human Sm-D antigen. Proc Natl Acad Sci USA 85:4832
25. Yamamoto K, Miura H, Moroi Y (1988) Isolation and characterization of a complementary DNA expressing human U1 small nuclear ribonucleoprotein C polypeptide. J Immunol 140:311
26. Habets WJ, Sillekkens PTG, Hoet M (1987) Analysis of a cDNA clone expressing a human autoimmune antigen: full length sequence of the U2 small nuclear RNA-associated B″ antigen. Proc Natl Acad Sci USA 84:2421
27. Darwin BS, Grudier JP, Klatt CL (1986) IgG antinuclear antibodies with cross-reactive rheumatoid factor activity. J Immunol 137:3796
28. Pisetsky DA, Caster SA (1984) Idiotypic cross-reaction between MRL autoantibodies with different antigen specificity. Clin Immunol Immunopathol 30:461
29. Migliorini P, Ardman B, Kaburaki J et al. (1987) Parallel sets of autoantibodies in MRL-*lpr*/*lpr* mice. An anti-DNA, anti-SmRNP, anti-gp70 network. J Exp Med 165:483
30. Kaburaki J, Stollar BD (1987) Identification of human anti-DNA, anti-RNP, anti-Sm, and anti-SS-A serum antibodies bearing the cross-reactive 16/6 idiotype. J Immunol 139:385
31. Provost TT, Watson R, Gammon WR (1987) The neonatal lupus syndrome associated with U1 RNP (nRNP) antibodies. N Engl J Med 316:1135
32. LeFeber WP, Norris DA, Ryan SR (1984) Ultraviolet light induces binding of antibodies to selected nuclear antigens on cultured human keratinocytes. J Clin Invest 74:1545
33. Eisenberg RA, Dyer K, Craven SY (1985) Subclass restriction and polyclonality of the systemic lupus erythematosus marker antibody anti-Sm. J Clin Invest 75:1270
34. Eisenberg RA, Craven SY, Cohen PL (1987) Isotype progression and clonality of anti-Sm autoantibodies in MRL-*lpr*/*lpr* mice. J Immunol 139:728
35. Tan EM (1982) Autoantibodies to nuclear antigens: their immunobiology and medicine. Adv Immunol 33:167
36. Shores EW, Eisenberg RA, Cohen PL (1986) Role of Sm antigen in the generation of anti-Sm autoantibodies in the SLE-prone MRL mouse. J Immunol 136:3662
37. Reuter R, Luhrmann R (1986) Immunization of mice with purified U1 small nuclear ribonucleoprotein (RNP) induces a pattern of antibody specificities characteristic of the anti-Sm and anti-RNP autoimmune response of patients with lupus erythematosus, as measured by monoclonal antibodies. Proc Natl Acad Sci USA 83:8689
38. Query CC, Keene JD (1987) A human autoimmune protein associated with U1 RNA contains a region of homology that is cross-reactive with retroviral $p30^{gag}$ antigen. Cell 51:211
39. Weening JJ, Fleurin GJ, Hoedemaeker PJ (1978) Demonstration of antinuclear antibodies in mercuric chloride-induced glomerulopathy in the rat. Lab Invest 39:405
40. Reuter R, Tessars G, Vohr HW et al. (1989) Mercuric chloride induces autoantibodies against U3 small nuclear ribonucleoprotein in susceptible mice. Proc Natl Acad Sci USA 86:237
41. Pelletier L, Pasquier R, Hirsch F (1986) Autoreactive T cells in mercury-induced autoimmune disease: in vitro demonstration. J Immunol 137:2548
42. Oliner H, Schwartz RS, Dameshek W (1961) Studies in experimental autoimmune disorders. I. Clinical and laboratory features of autoimmunization (runt disease) in mice. Blood 17:20
43. Gelpi C, Rodriguez-Sanchez JL, Martinez MA (1988) Murine graft versus host disease. A model for study of mechanism that generate autoantibodies to ribonucleoproteins. J Immunol 140:4160

44. Datta SK, Schwartz RS (1976) Autoimmunization and graft versus host reactions. Transpl Rev 31:44
45. Rolnik AG, Pals S, Gleichmann E (1983) Allosuppressor and allohelper T cellks in acute and chronic graft versus host disease. II. F1 recipients carrying mutations at H-2K and/or I-A. J Exp Med 157:755
46. Zouali M, Stollar BD, Schwartz RS (1988) Origin and diversification of anti-DNA antibodies. Immunol Rev 105:137
47. Ropes MW, Bauer W (1964) Observations on the natural course of disseminated lupus erythematosus. Medicine 43:387
48. Eisenberg RA, Craven SY, Warren RW (1988) Stochastic control of anti-Sm autoantibodies in MRL-*lpr*/*lpr* mice. J Clin Invest 80:691
49. Cooke A, Lydyard P, Roitt I (1984) Autoimmunity and idiotypes. Lancet ii:723
50. Wasserman NH, Penn AS, Freimuth PI (1982) Anti-idiotypic route to anti-acetylcholine receptor antibodies and experimental myasthenia gravis. Proc Natl Acad Sci USA 79:4810
51. Shechter Y, Maron R, Elias D (1982) Autoantibodies to insulin receptor spontaneously develop as anti-idiotypes in mice immunized with insulin. Science 216:542
52. Mendlovic S, Brocke S, Shoenfeld Y et al. (1988) Induction of a systemic lupus erythematosus-like disease in mice by a common human anti-DNA idiotype. Proc Natl Acad Sci USA 85:2260
53. Ollier P, Rocca-Serra J, Somme G (1985) The idiotypic network and the internal image: possible regulation of a germ-line network by paucigene encoded Ab2 (anti-idiotypic) antibodies in the GAT system. EMBO J 4:3681
54. Bona CA, Finley S, Waters S (1982) Anti-immunoglobulin antibodies. III. Properties of sequential anti-idiotypic antibodies to heterologous anti-gamma globulins. Detection of reactivity of anti-idiotype antibodies with epitopes of Fc fragments. J Exp Med 156:986
55. Chen PP, Fong S, Houten RA (1985) Characterization of an epibody. An anti-idiotype that reacts with both the idiotype of rheumatoid factors and the antigen recognized by rheumatoid factor. J Exp Med 161:323
56. Bruck C, Co, MA, Slaoui M (1986) Nucleic acid sequence of an internal image-bearing monoclonal anti-idiotype and its comparison to the sequence of the external antigen. Proc Natl Acad Sci USA 83:6578

Chapter 11

Inappropriate HLA Molecule Expression in Epithelial Cells: Relevance for Human Autoimmunity

R. Mirakian, A. Ciampolillo, U. Deuss, A. Miyazaki and G. F. Bottazzo

Introduction

The central dogma of immunology, originally postulated by Paul Ehrlich, is based on the assumption that the immune system does not normally react to "self". Consequently, autoimmunity represents the state in which the natural tolerance to self tissue and cell components terminates.

Although the molecular basis of antigen recognition in humoral immunity (i.e. antibody production) has long been understood, the mechanism by which immunoregulatory and cytotoxic T (Tc) lymphocytes recognize antigen have only recently been clarified. Of particular importance has been the characterization of the T cell receptor (TCR) for antigen [1] and the demonstration that T cells, unlike B lymphocytes, predominantly recognize antigen in association with membrane-bound products of the major histocompatibility complex (MHC) [2,3]. This model is known as the "genetic" or "MHC restriction phenomenon", and has been found to be operative in many species, including humans (for review see [4]).

The class I and class II MHC molecules are transmembrane glycoproteins encoded by the genes of the HLA-A, -B and -C and the HLA-DR, -DQ and -DP subregions of the human MHC complex, respectively [5]. HLA class I molecules are present in most nucleated cells and are a prerequisite for the recognition of antigens by Tc ([6] and review in [7]). The expression of HLA class II molecules is usually restricted to cells of the immune system, such as macrophages, dendritic cells, B lymphocytes and activated T cells, and to some capillary endothelial cells. They play a key role in the presentation of antigens and in the regulation of the immune response. The activation of helper T (Th) lymphocytes requires the simultaneous interaction of the T cell receptor with the antigen/class II molecules complex on the membrane of the presenting cell [8]. In a physiological immune response, where the macrophage functions as antigen-presenting cell (APC), the antigen (e.g. a polypeptide) is first processed by partial digestion and selected portions of the molecule are reinserted in the cell membrane, where they then form tight complexes with class II molecules for their subsequent presentation to Th cells.

The efficiency of antigen presentation depends on the stability of the complexes formed by the antigen and the class II molecule. The allelic variations in the corresponding gene coding regions may therefore explain the diversity and potency of immune responses among different individuals in a given species. In addition, tolerance to highly specialized cells, like endocrine cells, can be maintained by their "immunological silence", i.e. these cells do not express HLA class II molecules on the cell surface, thus "ignored" by the Th which can recognize them only when presented in the context of class II molecules. However, this state of "immunological silence" in certain circumstances can be broken, thus creating the ideal conditions for immunocytes to recognize self molecules (e.g. autoantigens) which they, apparently, have never "seen" before and toward which they are not tolerised. If this did occur, the net result would be the triggering of an autoimmune response with inevitably deleterious consequences for the cells in which the phenomenon has taken place (for review see [9]).

HLA class II molecules were assumed to be expressed only in organs of the "classical" immune system, but it is now clear that most of the human tissues contain a variety of class II positive cells [10–13].

Cells with these characteristics are, for example, those with a dendritic morphology, such as the Langerhans cells of the skin or the Langerhans-like cells of the cornea, and cells akin to macrophages, such as the Kupffer cells of the liver or the alveolar macrophages. There are epithelial cells, which physiologically express class II products. This refers particularly to cells lining cavities, as those in the respiratory or gastrointestinal tracts, constantly exposed to environmental antigens. The precise role of HLA class II expression in these cells is still under active investigation but, among other functions, it has been shown that they might act as accessory APCs to an unusual subset of suppressor T (Ts) lymphocytes [14].

In addition, class II molecules have been detected on epithelial cells of salivary glands (parotid and submandibulary), gall bladder, lactating mammary glands, urinary bladder and kidney pelvis [10–13]. In some individuals, class II molecules can be detected in tubular epithelial cells of the kidney. Interestingly, the intensity of class II expression in these cells has been suggested to influence the outcome of a kidney transplant [15,16]. In the pancreas, some epithelial cells of the small ducts are also class II positive [17], as confirmed by immunoprecipitation studies with isolated pancreatic cells [18]. If practically all endocrine cells are class II negative, the zona reticularis of the adrenal cortex does not follow the expected rule. In fact, a frequent occurrence of class II positive epithelial cells in this area has been reported in normal glands [19,20].

A moderate-to-intense expression of HLA class II molecules can be found frequently in capillary endothelial cells. This has been shown, for example, in the glomerular and peritubular endothelium of the kidney [12] and in a proportion of endothelial cells in the human thyroid (for review see [21]). In the normal pancreas, the fine network of capillary endothelial cells are almost invariably positive for class II, both in the exocrine portion as well as around and inside islets [22]. An exception seems to be represented by capillary cells in the normal human heart which are found constantly negative for class II [23].

In summary, strong class II expression is restricted to immunocompetent cells. A moderate expression is present in some epithelial cells which are exposed to the constant entry of environmental antigens, whereas epithelial cells in less vulnerable tissues, like the endocrine glands, are generally class II negative.

"Inappropriate" Expression of HLA Class II Molecules in Organ-Specific Autoimmune Diseases

The first evidence appeared when it was shown that normal thyroid cells when cultured in monolayers could be induced by lectins to express HLA class II molecules in vitro [24]. Most importantly, an inappropriate expression of the same molecules was detected in tissues affected by thyroid autoimmune disease [25]. When the analysis was extended to frozen sections of several surgical thyroid specimens, HLA-DR expression was most marked in Hashimoto's glands, when all surviving follicles were strongly stained by the corresponding monoclonal antibodies (MoAbs). In the majority of thyroids from patients with Graves' disease, the phenomenon was seen in discrete groups of follicles but the intensity of the reaction was also quite pronounced. In addition, a positive staining was also detected, but in a patchy distribution, in some non-toxic nodular goitres, whereas it was constantly absent in normal control glands. These results have subsequently been confirmed on a large sample of thyroids from a variety of pathologies, including thyroid tumours [26–30].

With the exception of the latter [26], and to a lesser extent in Graves' thyroids [25,26], class II on thyroid follicular cells has always been found in the context of an active lymphocytic infiltration or presence of focal thyroiditis [27–30]. Interestingly, the Graves' glands that were negative for class II staining came from patients without circulating thyroid antibodies. In contrast, class II expression detected on thyroid surgical specimens from patients affected by a variety of thyroid diseases almost invariably correlated with the titres of circulating thyroid antibodies [26]. When the study was extended to the MHC subregion distribution of inappropriately expressed HLA class II products on the same autoimmune thyroids, this followed the pattern DR>DP>>DQ [31], similarly to that observed in APCs. There was a positive correlation between the presence and intensity of DR expression and high titres of antibodies to thyroid microsomal antigen, now identified as the thyroid peroxidase (TPO) [32–35], and a similar relationship was recorded between the expression of DQ and high titre of anti-thyroglobulin antibodies [36]. Interestingly, the production of anti-salivary antibodies in Sjögren's syndrome, a condition in which the duct epithelium in the gland also inappropriately expresses class II [37], is linked to HLA-DQ subregion genes [38].

The unique chance to obtain a freshly frozen pancreas of a girl who died of ketoacidotic coma soon after diagnosis, gave the first opportunity to study the phenomenon of inappropriate expression of HLA class II molecules in the β cells of the islets of Langerhans in type I diabetes mellitus [22]. Remarkably, class II expression was restricted to insulin-containing islets, and by double IFL with the corresponding anti-hormone antisera it was conclusively proved that only β cells inappropriately expressed HLA class II molecules, whereas glucagon, somatostatin and exocrine cells were negative. In contrast to the thyroid, class II positive β cells were predominantly seen in islets free of lymphocytic infiltration. Class II expression was also significantly increased in capillary endothelial cells around diseased islets which were unusually hypertrophied and dilated, a phenomenon that was not observed in the capillary network of the exocrine portion of the gland. Expression of HLA class I molecules, which is weak on normal pancreatic endocrine cells, was markedly increased in all the cells within the islet, but not in the exocrine cells.

Soon after these findings were reported, similar observations were confirmed following the investigation of several fixed pancreas blocks obtained from autopsies of type I diabetic patients who died at or soon after diagnosis of diabetes [39]. After examining around 3000 islets it was shown, once again, that only islets containing insulin positive β cells inappropriately expressed HLA class II molecules and, unexpectedly, in about 70% of these islets there was no lymphocytic infiltration. The lack of an active insulitis process in these islets was conclusively demonstrated by cutting at least 40 serial sections and staining them with specific MoAbs to leukocytes, thus excluding interior trapping of inflammatory cells in the islets containing class II positive β cells. By extrapolation, such islets may depict what the surrounding insulin-deprived islets would have looked like many years previously when the entire process was initially triggered. Hyperexpression of MHC class I molecules on the islets and hypertrophy of class II positive islet capillary endothelial cells were also confirmed, and the observation extended to several blocks of pancreas from newly diagnosed type I cases [40]. β cells in islets of type II, or non-insulin dependent, diabetic patients were continuously negative for class II and, remarkably, class II expression in pancreatic β cells has never been detected in other inflammatory diseases of the pancreas, such as cystic fibrosis or chronic pancreatitis, despite the presence of heavy inflammatory infiltrates surrounding the islets [17,39].

Since the initial description of inappropriate HLA class II expression in thyroid autoimmune disease and in type I diabetes, several groups, including our own, have detected similar phenomena in a wide range of tissues undergoing autoimmune attacks (for review see [41]). The latest acquisitions to the list are represented by the autoimmune variant of the protracted diarrhoea in children [42], chronic active hepatitis [43] and fibrosis alveolitis [44]. This extended evidence has consolidated the original concept and the presence of MHC products in epithelial cells is now generally considered an important addition to the known pathological characteristics of organ- and, possibly, non-organ specific autoimmunity.

Can Class II Positive Epithelial/Endocrine Cells Function as Classical Antigen-Presenting Cells?

Our hypothesis that endocrine autoimmunity may be "controlled" by the ability of non-classical APCs to present their own autoantigen to Th cells [45] has been substantiated by several observations. T lymphocytes, for example, can indeed recognize class II positive epithelial cells after co-culture with autologous, PHA stimulated, normal thyrocytes [46]. A T cell clone specific for a defined peptide fragment (p20) of the haemagglutinin molecule of the influenza virus, was stimulated by Graves' HLA class II positive thyrocytes of the appropriate HLA phenotype in the presence of p20 [47]. No stimulation was observed with a distinct peptide of the same haemagglutinin. In addition, these thyrocytes were unable to present the whole virus, alive or dead. This latter finding is in agreement with the fact that epithelial cells, in contrast to APCs, cannot process complex antigens for presentation at the cell surface. However, antigen processing is not required in an autoimmune situation as epithelial cells bear their own surface autoantigens in association with class II molecules already as "processed" products inserted in their plasma membrane, where they could then be recognized by Th cells.

Although relevant, this experimental data did not resolve the decisive question of whether thyrocytes expressing class II could efficiently present self surface constituents, i.e. relevant autoantigens. This was achieved through cloning autoreactive Th cells selected from activated lymphocytes infiltrating Graves' thyroids [48]. It was shown that certain T cell clones could indeed proliferate after exposure to autologous class II positive thyrocytes. In contrast, allogeneic class II positive thyrocytes isolated from other Graves' patients, or autologous peripheral blood mononuclear cells (PBMC) from the same Graves' patient did not exert any effect, thus confirming the MHC restricted nature of the recognition and the epithelial cell specificity of the phenomenon. Importantly, the stimulation of autoreactive Th cell clones by autologous class II positive thyrocytes was blocked by MoAbs to class II. Subsequent experiments from other laboratories have substantiated these initial findings [49–51]. All these data are certainly relevant to autoimmunity because they could not be reproduced in multinodular non-toxic colloid goitres, a disease in which inappropriate expression of HLA class II is not a rare event [52]. Thyrocytes derived from these glands were unable to operate as APCs, since lines obtained from lymphocytes infiltrating the goitre did not succeed in recognizing autologous thyrocytes. This strongly suggests that the antigen-presenting capacity and class II expression do not necessarily lead to the generation of autoreactive T cells.

It is now accepted that the initiation of a specific immune response requires not only an antigen-dependent capacity but also antigen-independent interactions which appear to be driven by recently identified accessory molecules (for reviews see [53,54]). It has, for example, been demonstrated through transfection experiments that the co-expression of ICAM-1 in mouse L cell (a fibroblastoid cell line) transfectants which expressed a modest level of class II molecules was critical for an effective class II restricted T cell recognition and their subsequent activation [55]. Class II alone on the transfected fibroblasts was insufficient for antigen presentation to T cell clones specific for a given viral peptide. Adhesion molecule expression on epithelial/endocrine cells affected by autoimmunity has not yet been investigated, but work in this direction, using normal endocrine cells in cultures, has now been initiated [56,57]. Interestingly, new information has been produced in autoimmune protracted diarrhoea of infancy [58], where affected enterocytes not only inappropriately express class II but also ICAM-1 [59].

Are There Any Lessons From Transgenic Mice?

If the early and the mid 1980s witnessed a succession of reports describing inappropriate expression of HLA class II products in epithelial cells affected by autoimmunity, in the late 1980s efforts were directed toward evaluation of the possible pathogenetic role of the phenomenon. For this purpose, several investigators adopted the transgenic mouse model with the specific and direct aim of evaluating whether the introduction and subsequent expression of different MHC genes in epithelial cells in vivo could abrogate tolerance against the relevant autoantigens, with the subsequent appearance of infiltrating lymphocytes, causing ultimately immune destruction of the target cell. In other words, was it possible in the genetically manipulated animals to reproduce the same histopathological pattern observed in the human disease? It is not surprising that type I diabetes was the chosen model in these experiments.

Initially, three groups prepared constructs with different cloned H-2 genes using insulin gene promoters from different species. These constructs were then repeatedly injected into fertilized mouse oocytes to generate transgenes. Allison et al. manipulated class I H-$2K^b$ genes [60], Lo et al. class II I-E^b genes [61], and Sarvetnich et al. class II-A^d genes [62]. Only in the last study was a syngeneic approach adopted. The results were surprisingly similar in the three groups of experiments. In all cases, a clear class I or class II transgene expression was found in pancreatic β cells, and the mice undoubtedly became frankly diabetic. Unexpectedly, there was no evidence of lymphocytic infiltration in association with high degrees of β cell dysfunction. Neonatal thymectomy did not affect the course of the disease [60], thus excluding a prominent role for T cells in its causation. In addition, no response could be evoked in vitro by peripheral or thymic T cells to allogeneic Ia (e.g. I-E^b) gene products, although a good response to other class II haplotypes was maintained [61]. This clearly indicated that the transgenes were in fact made tolerant to the "foreign" introduced MHC product and suggested that this might be due to "homing" of "stimulated" peripheral T cells back to the thymus [62].

In all three models a considerable reduction in insulin secretion by the islets was documented, despite wide differences in the morphological appearance and degranulation of the β cells. It was postulated that the most likely explanation for this type of diabetes in the transgenic mice was the disruption and alteration of the cytoplasmic pathways involved in insulin synthesis [63], possibly resulting in perturbation of the intrinsic secretory machinery. In other words, insulin itself or other auxiliary peptides, which are required physiologically for hormone biosynthesis and subsequent exocytosis, could become "trapped" into the "groove" recently identified on the MHC molecule [64], thus forming a tight complex with the MHC molecule, themselves concomitantly biosynthesized in the cytoplasm. Obviously, if such a process did occur, it would severely restrict the secretion of the hormone with subsequent profound impairment of its peripheral action.

More recently, different results were obtained using a model of transgenes derived by the injection of allogeneic I-$A^k{}_\alpha$ I-$A^k{}_\beta$ cDNA constructs under the control of a rat insulin promoter [65]. In these experiments, MHC transgene transcripts were detected on the surface of the pancreatic islet β cell at a much lower level of expression than in the previously produced transgenic systems. By aiming for a physiological expression of class II products on the β cells (similar to that detected on APCs), transgenic positive littermates showed no signs of overt diabetes. However, as previously, no lymphocytic infiltration was produced around islets but, surprisingly, the immunological "inactivity" did not result from tolerance to the "foreign" MHC class II molecules, since MLR experiments demonstrated that most lymphocytes could respond to the transgenic I-A.

Diverging for a moment, it is important to mention that in the first transgenic mouse report [62] there was also the description of similar mice made transgenic with an interferon (IFN-γ) gene, whose product was again expressed by β cells in the islets. Even these animals became overt diabetic but, in contrast to the previously described diabetic MHC transgenic models, this time their pancreas showed massive mononuclear cell infiltration. However, the histopathological appearance was that of a pattern of generalized inflammation which was quite different from that observed classically in the human insulitis. In fact, not only the islets were inflamed but also the exocrine tissue, resembling, at a closer view, the pattern detected in pancreatitis rather than that seen in type I diabetes.

Returning to the MHC transgenic models, it has been speculated [65] that the unexpected failure to produce overt diabetes and breakdown of tolerance to unique surface β cells autoantigens could be the consequence of several events: (a) the "timing" of class II appearance on cells, which are normally devoid of these molecules, could be relevant, i.e. the phenomenon becomes "truly" inappropriate when it occurs later in life, rather than during fetal development or close to birth; (b) the amount of MHC products synthesized in the islet β cells is certainly critical and, whereas low levels of expression appear not to hinder insulin synthesis/secretion, when these molecules are produced in abundance they apparently compete for binding of intracellular peptides, including insulin. One can further expand speculations and suggest that the initial design of the construct is of fundamental importance. For example, expression of only one allele of the class II gene products, despite the presence of both α and β chains, may not be sufficient for an efficient autoantigen presentation. In addition, one should not forget the essential role of accessory molecules in antigen presentation, e.g. cells were able to present relevant peptides to specific T cell clones only when ICAM-1 [55] or the invariant chain [66] were concomitantly synthesized and expressed on these cells.

Based on these considerations, the lesson emerging from these transgenic models is, to say the least, very unclear. So far, none of the models produced an active insulitis, thus making this experimental system very different from the human type I diabetes (for review see [67]). Moreover, the majority of the transgenic mice have offered the opportunity to investigate an "allo" rather than an "iso" situation, possibly offering new insights for problems related to transplantation. Paradoxically, when class II transgenic islets were transplanted into MHC incompatible mice, they were not rejected, as one might have initially predicted [68]. Even more surprising was the concomitant demonstration that, when used as potential APCs in an in vitro system, class II positive islets suppressed rather than stimulated T cell clones able to respond to a given viral antigen when presented by conventional APCs [68]. This might indicate that, in certain circumstances, partial expression of class II products may have a potentially beneficial effect.

In summary, these particular transgenic mice certainly did not help to explain the role of aberrant MHC product expression in islets of type I diabetic patients or in other tissues affected by autoimmunity. However, the models could provide new clues on the mechanisms involved in hormone synthesis and exocytosis, particularly in relation to the effect of MHC molecules in the intracellular processes. In addition, the model might contribute to a broader understanding of the present immunological "difficulties", such as those relating to thymic education and maintenance of tolerance at the level of the T cell compartment. Regarding autoimmunity, it is quite clear that these particular types of genetic manipulations are still in their infancy and further but decisive refinements are necessary for the model to become useful. Only when these aims are achieved will transgenic mice take their rightful place among the presently available animal models of autoimmunity.

How is MHC Molecule Expression in Endocrine Cells Triggered and Regulated?

Are Cytokines Responsible?

IFN-γ, but not IFN-α, IFN-β or interleukin-2 (IL-2), is the potent lymphokine that can induce or enhance class II expression in many cells of non-lymphoid origin in vitro (for review see [69]), including thyrocytes [70], keratinocytes [71], and melanocytes [72]. On the other hand, mediators which inhibit class II expression have also been identified, such as α-fetoprotein, on macrophages [73]. Therefore, it is likely that in the in vivo situation a complex network of immune mediators must exist and be responsible for giving positive as well as negative signals to cells with a predetermined response pattern, such as those involved in a temporal regulation of class II molecule expression. This is in agreement with the current view that MHC class II gene transcription is influenced by several regulatory sequences, some IFN-γ dependent [74] and some independent [75].

Because of the ease of availability of tissue, the regulation of class II molecule expression has been most intensively studied in vitro in the human thyroid system. It has been shown that IFN-γ induces class II expression in nearly 100% of normal thyroid cells cultured in monolayers [49,70] and that HLA class II subregion expression parallels again the hierarchy of expression normally detected on APCs. Low doses of IFN-γ (5–10 Uml^{-1}) induced strong expression of DR but very little DP or DQ, whereas DP and DQ are induced in a high proportion of thyrocytes (though at a lower level than DR) when a higher dose of IFN-γ (500 Uml^{-1}) was employed [31]. As mentioned, the same order of intensity, i.e. DR>DP>DQ, was observed in thyroid follicular cells of glands affected by autoimmunity, when sections were stained with the corresponding MoAbs. TSH and thyroid stimulating immunoglobulins (TSI) also enhance thyrocyte class II expression and they are most effective when used in combination with a suboptimal dose of IFN-γ [76,77]. This finding may explain the strong class II staining detected in thyrocytes of patients affected by Hashimoto's thyroiditis, a disease known to be accompanied by an elevated level of circulating TSH. As the enhancing effect of TSH was mimicked by dibutyryl-cyclic-AMP (db-cAMP), one can extrapolate that cAMP is the second messenger for this activity, as it is for many effects of TSH on thyrocytes. When thyrocytes were incubated with IFN-γ and TNF-α, a synergistic effect of the two cytokines on class

II expression was observed, which is accounted for by the upregulation of IFN-γ receptors by TNF [78].

With regard to factors that negatively affect the expression of class II molecules, it has been shown that epidermal growth factor (EGF), which has some counter-regulatory action for TSH, can suppress by 50% the class II expression induced in thyrocytes after incubation with IFN-γ [79]. In addition, Transforming growth factor-α (TGF-α) [79] and IFN-α_1 [80] also inhibit the induction of class II expression on thyrocytes, an effect, at least in the latter case, attributable to receptor interference/regulation with IFN-γ.

In contrast to thyrocytes where it is relatively easy to induce and regulate HLA class II expression with cytokines or other stimuli, the situation with human pancreatic islet cells is much more complex. A panel of crude and purified preparations of cytokines, including IFN-γ, a variety of lectins and hormones, as well as chemicals, known to directly stimulate cAMP and calcium-dependent transduction pathways, were found to induce the expression of HLA class II molecules in the majority of pancreatic cell types present in monolayers. Islet endocrine cells, however, were totally refractory to these potential inducers, and remained constantly negative for class II [81]. Certainly IFN-γ caused hyperexpression of class I molecules on islet cells, but it was only IFN-γ with TNF-α or lymphotoxin (TNF-β) known to have synergistic effects in several experimental systems [82], that produced class II induction in cultured human islet cells [83]. Despite numerous trials, this still remains the only workable cytokine combination. In fact, IFN-γ with any other cytokines or substances proved to be ineffective. However, it is important to emphasize that the same combination of cytokines exerted a similar effect also on glucagon and somatostatin cells which, as repeatedly shown, are devoid of class II in "diabetic" islets in vivo. Class II molecules appear exclusively on the target β cells in diabetic insulitis [22,39].

Is Inappropriate HLA Class II Expression in Endocrine Cells a Primary or a Secondary Event?

At this point of the discussion, the above question becomes crucial. After a critical evaluation of the facts discussed above, we favour the first possibility, i.e. the endocrine cells have the intrinsic potential to trigger the phenomenon independently of its positive response to the action of immune factors. One admits that, when autoimmune thyroid glands are examined, it is difficult to disentangle the actual temporal sequence of events occurring in vivo, since the inappropriate expression of class II molecules seems generally to coincide with the lymphocytic infiltration. However, this correlation is not so tight in Graves' thyroids and in certain thyroid carcinoma where class II expression on follicular cells is found in the absence of active inflammation, when proper morphometrical analysis is carried out [26].

If we apply similar considerations to the islets, the situation appears quite different and much clearer than in the thyroid. First, if one recalls the results obtained in the diabetic pancreas, where in different islets several steps of the autoimmune process are visible at the same time, the majority of β cells express class II molecules in the absence of insulitis [39,40]. In two-thirds of the islets with aberrant class II expression on β cells, no lymphocytes could be detected. This observation makes it unlikely that class II expression is primarily induced in vivo by immunomediators secreted by invading lymphocytes. Secondly, one may recall that in chronic pancreatitis or cystic fibrosis, β cells remain totally class II negative despite the abundant presence of inflammatory mononuclear cells in the exocrine compartment [17,40]. Furthermore, in segmental pancreatic grafts transplanted to type I diabetic twins from their non-diabetic identical co-twins, β cells in the transplanted pancreas remained class II negative despite direct evidence of lymphocytic infiltration surrounding the islets, as a direct result of the recurrency of insulitis in these twins ([84] and see below). Clearly, in all these situations islets must be exposed to a fairly large amount of cytokines and, despite that, these cells remain class II negative.

The message that has emerged from these extensive studies is that cytokines lack β cell specificity for the induction of class II expression in vitro and it is unlikely they can act as primary effector mechanism in vivo, at least in type I diabetes. One cannot exclude, however, that their synergistic effect may be important for the perpetuation of class II expression, once triggered on epithelial/endocrine cells.

What Role do Cytokines Play in Endocrine Autoimmunity in General?

Another important aspect to consider is the distinct action that cytokines exert in vitro when compared with their in vivo effects. Studies in vitro, although instructive, have obvious limitations, as the following two examples demonstrate. When the mode of action of TNF on cultured endothelial cells isolated from umbilical cords was studied, a profound inhibiting action on a variety of cell functions, including proliferation, was observed [85]. In contrast, when the same cytokine was injected in vivo it enhanced proliferation of endothelial cells, thus exerting exactly the opposite effect [86–88]. The second example illustrates that the intrinsic status of a cell may determine its response to a cytokine. Normal resting thyroid cells and thyrotoxic thyrocytes (regardless whether they are obtained from Graves' thyroids or "hot" nodules) behave differently when exposed to IL-1α ([89] and unpublished work). In the former situation, both proliferation and cAMP stimulation by TSH are depressed in the presence of the cytokine, whereas constitutively hyperstimulated thyrocytes have the reverse response to it. Clearly, these two examples emphasize how cautious one must be in extrapolating information obtained from in vitro studies to the interpretation of in vivo events.

Extrapolation is even more dangerous when considering the effects of cytokines in response to common and frequent hyperreactive situations (for review see [90]). These include: stress, cachexia, febrile status caused by generalized viral or other noxious infections, destructive or stimulating autoimmunity, and cancer. In addition, it is now clear that cytokines, far from reacting as a separate entity, participate in important "communicating" networks in the body. They are known to operate in close communication with the hypothalamus/pituitary/adrenal axes (see reviews in [91,92]). The importance of this complex endocrine system in regulating physiological immune responses is well established and the part it may play in influencing autoimmunity has often been suggested. But what about the interaction of cytokines with sex hormone producing cells? There is evidence that this might also be the case, at least from data emerging from animal studies [93,94]. After all, it is well known that classical endocrine autoimmunity affects females more commonly than males and recent evidence suggests that cytokine treatment in cancer patients has caused the first appearance or worsening of subclinical thyroid autoimmunity more readily in the treated female patients ([95–97] and review in [98]).

Cytokines can also act as important "killer" factors in vitro, when exposed to certain endocrine cells. IL-1α exerts interesting differential effects, being cytostatic for thyroid cells (for review see [99]) but predominantly cytotoxic for islet cells (for review see [100]). IFN-γ and TNF-α in combination, are readily cytotoxic for both types of cells when incubated for a prolonged period on monolayer cultures [101,102]. However, when one considers the proposed aetiological role for IL-1 as *primum movens* for damaging β cells, this concept has a drawback when reflecting that macrophages (the cells that produce the cytokine) are very rarely present in human "diabetic" islets [39]. Thus, we have again demonstrations of actions in vitro which cannot readily be extrapolated to the proposed pathogenetic effect of cytokines in vivo or to the actual appearance of the pathology in humans, as it has been recently emphasized in a lively debate (see "correspondence" which followed [100]).

Let us consider another aspect, the demonstration that endocrine cells themselves can indeed produce cytokines. In type I diabetes, β cells still containing insulin also contain IFN-α [103], and cultured β-cells have been shown to synthesize and secrete IL-6 [104]. These findings open up competely different perspectives for future interpretations. On the one hand, they suggest that a latent noxious agent(s) may be present in the target cells and the switching on of the physiologically downregulated IFN-α gene could be interpreted as its response to the putative invading factor in the endocrine cell. On the other hand, endocrine cells could acquire other unexpected "immunological" properties, in accordance with the concept of their direct participation in the immunological circuit, and their ability to actively communicate with immunocytes and activated endothelial cells surrounding them. Incidentally, it has also been suggested that IFN-α, if secreted by "diabetic" β cells could have a paracrine effect and be responsible for the paradoxical enhanced expression of HLA class I molecules on glucagon and somatostatin cells (for review see [105]). Why these cells do not attract cytotoxic T lymphocytes, as do β cells with similar enhanced class I expression, remains unresolved. In this context, it is important to emphasize that in "diabetic" islets deprived of class II^+/IFN-α^+ β cells, glucagon and somatostatin cells retain a physiological expression of class I. Thus the phenomenon appears to be self-limiting in the cells that are not ultimately killed in type I diabetes but much more relevant for β cells which inevitably will be destroyed. Likewise for class II, it remains to be shown what factors are initially responsible for enhancing class I expression in the target cells.

In summary, the role of cytokines in autoimmunity is not yet framed in a clear and comprehensive picture. It is now imperative to interpret data in an integrated manner with more emphasis put on in vivo studies to avoid increasing confusion and erroneous extrapolations.

Is There a Role for Viruses in Human Endocrine Autoimmunity?

There are examples in which the infection or transfection with DNA [106] and RNA [107,108] viral proteins are sufficient to perturb MHC gene expression in certain cells or cell lines, both upregulation and downregulation of MHC products being observed [109–111]. Recently, transfection of cultured human thyroid cells with a mutant of the SV40 virus induced de novo expression of HLA class II molecule and enhanced class I expression ([112] and unpublished work). These lines and cloned thyrocytes, therefore, provide an unexpected new model for investigating how latent viral infection may switch on or modulate MHC products on epithelial cells. This, in turn, could initiate the chain of events leading to autoimmune responses, as originally hypothesized [48].

It has long been suspected that environmental agents (e.g. viruses) could be responsible for causing autoimmune diseases, a suspicion strongly supported in type I diabetes (for review see [113]. Repeatedly, common viruses (i.e. coxsackie, mumps, rubella, influenza, etc.), together with certain chemical compounds [14] and food additives ([115] and review in [116]), have been claimed to be involved in β cell damage and possibly other endocrine cells affected by autoimmunity. Nonetheless, direct evidence for their precise role in the killing mechanism is still lacking. It is well known that these viruses tend to cause rapid damage when exposed to certain cells in culture, including β cells. However, these data clash directly with the now accepted concept that the clinical onset of autoimmune diseases is preceded by a latency period of many years (for review see [117]). This is best exemplified in type I diabetes. In fact, genetically susceptible individuals found to be positive for ICA, e.g. first degree relatives of diabetic probands [118,119], identical twins [120] or triplets [121], polyendocrine patients [122], and even "sporadic" cases (i.e., cases with no family history of the disease [123]) can live asymptomatically for many years without showing any signs or symptoms of β cell dysfunction.

From this it follows that the previously incriminated "common" viruses or other toxic substances can be envisaged in the pathogenetic mechanism only as *precipitating* rather than *initiating* factors (for review see [124]). Presumably, the autoimmune damaging process initiated in earlier years renders the target cells prone to invading "common" viruses, which might ultimately kill them. However, unexpectedly, no evidence of coxsackie virus was reported in "diabetic" β cells, when a polyclonal antiserum to VP1 (the protein common to all coxsackie viruses and expressed in the virulent state) was applied on sections of pancreas of newly diagnosed type I diabetic children who died close to diagnosis (for review see [105]). Thus, these data cast serious doubts even on the above proposed "precipitating" concept, at least in regard to the coxsackie strains.

Clearly, new hypotheses are required if viruses are to be implicated. This is the reason why we were particularly attracted to the possible involvement of slow viruses in human autoimmunity. Apart from the long latency period preceding the clinical manifestations of autoimmune diseases in humans, the hypothesis found further support in the recently accumulated evidence of an enhanced expression of endogenous retroviruses in spontaneous experimental organ-specific autoimmunities, i.e. autoimmune thyroiditis in the obese chicken [125] and diabetes in non-obese diabetic [NOD] mice (for review see [126]).

To complement these conceptual changes, came the demonstration that DNA extracted from thyroids of Graves' patients gave a positive hybridization signal by Southern blotting with the 0.7 and 0.9 kb of the *gag* probe of the HIV-1 retrovirus [127]. So far, about 60% of Graves' glands tested have been positive for these signals but, interestingly, in only half of the patients were similar results obtained from the DNA extracted from their PBMC.

These observations support the concept of a novel human pathogen retrovirus, which apparently has sequence homology with the known HIV-1 retrovirus. These preliminary data have posed several intriguing questions, to some of which there are already feasible answers. We suggest that these findings might point to an exogenous rather than endogenous retrovirus, mainly because DNA extracted from other thyroid pathologies or unrelated human organs did not give positive signal when hybridized with the same probes [128]. Because of the known infiltration of immunocytes into Graves' thyroids, it was important to distinguish whether the signals were in the DNA of thyrocytes or on invading mononuclear cells. Recent experiments based on the enrichment of immunocytes in Graves' glands, now indicate that the HIV-1 *gag* probe is strongly hybridizing only the DNA of immunocytes. Probes related to HTLV-1 gave negative results, suggesting that a retrovirus belonging to the lentivirus family, such as HIV-1, may be involved [128]. Clearly, these findings need to be substantiated (active work is in progress at present to try to finally clone the putative novel retrovirus), before the issues become overcrowded with further speculation.

The Growing Role of Endothelial Cells in Autoimmunity

Another aspect which may be relevant to pathogenesis is the hypertrophy of capillary endothelial cells around "diabetic" islets [22,39] and in other endocrine tissues affected by autoimmunity [25]. In addition to the hypertrophy, these capillaries strongly express both class I and class II molecules. Similar observations have been made recently in the heart where class II negative endothelial cells become strongly positive in pathological conditions suspected to have an autoimmune origin, such as dilated cardiomyopathy [23]. This phenomenon is at present under active investigation, but these structures could represent the "openings" which allow the subsequent invasion of autoreactive immunocytes. This process could be facilitated by an enhanced expression of adhesion and homing molecules, known to be actively synthesized by endothelial cells [129]. Such a "capillary/lymphocyte recognition system" has been revealed in rheumatoid arthritis [130].

Apart from the "homing" effect, capillary endothelial cells could also become involved in actively present autoantigens, as they appear to have this capacity for antigens in general (for review see [131]). Alternatively, they can produce and secrete factors which may contribute to the inappropriate expression of MHC products on the target epithelial/endocrine cells. Conversely, the target cell itself could contribute to the fine interaction with the surrounding capillary cells. This possibility is supported by the demonstration that porcine thyroid follicles secrete a heparin-binding growth factor which has a potential paracrine effect on adjacent endothelial cells [132]. Whether this factor indeed plays a role in human thyroid autoimmunity is not yet known. Future work will certainly bring new exciting insights into the endothelial/epithelial cell communication network, and its potential relevance for the pathogenesis of autoimmunity.

Is There a Decisive Role for Cytotoxic T Cells?

What about factors responsible for closing the "vicious circle"? External agents (? retroviruses, ? others), leukocyte products, or both, could also act in enhancing HLA class I expression, another inappropriate MHC phenomenon detected in "diabetic" islets (for review see [133]). This latter observation might now explain the abundant presence of Tc cells among infiltrating lymphocytes at the time of diagnosis [22]. These cytotoxic T cells appear to play a key role in the final selective destruction of β cells. The "accidental" in vivo experiment of the pancreas transplanted diabetic identical twins unexpectedly showed that diabetes recurred and insulin treatment was required after a few weeks in the transplanted diabetic twin. β cells alone were destroyed, and Tc cells once again were the predominant lymphocyte subpopulation present in the inflammatory infiltrate (for review see [84]).

These observations indicate that type I diabetic patients maintain memory Tc cells for several years and these can be "re-awakened" when class I identical β cells are introduced in the diabetic milieu. In addition, the pancreatic transplant experiments have shown that lymphocytes can cause rapid and irreversible decompensation of β cell function, in just a few weeks, thus indicating that when Tc cells are generated in sufficient numbers against β cells they do not leave much "breathing space" to these cells. This observation has important implications in relation to the common occurrence of ICA for long periods preceding overt clinical manifestation of type I diabetes. ICA, when produced, are apparently unable to cause gross damage to β cells; they can persist for several years in the blood without great interference with β cell function. It is only when Tc cells with fine specificity for β cells are generated that the "killing circle" is completed.

Inappropriate Expression of HLA Class II Molecules in Endocrine Cells: Reviving our Model for Autoimmunity

In the light of previous and new evidence presented in this review, we would like to conclude by updating our originally proposed model for the pathogenesis of endocrine autoimmune diseases [45].

We now postulate that viruses of the slow-replicating family (possibly retroviruses) may operate directly or indirectly at the level of three different types of cells: the target epithelial cells, the endothelial cells and the immunocytes. The facts currently confronting us are: (a) β cells in "diabetic" islets selectively express class II molecules and are, apparently, devoid of lymphocytic infiltration; (b) endothelial cells, which are hypertrophied, secrete cytokines (e.g. IL-1-β [134]) and hyperexpress MHC molecules. These abnormalities seem to be independent of the state of the endocrine cells, e.g. islets with normal appearance and with still large numbers of insulin cells or already atrophied with no insulin content; (c) the identification of retrovirus-like sequences in the DNA of immunocytes invading autoimmune thyroid glands.

At present, we have no knowledge of the precise temporal sequence in which these phenomena appear in the three cell types but "activation" of capillary cells appears a likely first event. It is true that we are totally ignorant of the factors responsible for causing the observed abnormalities in the endocrine and capillary cells. However, we cannot disregard the fact that "diabetic" β cells produce IFN-α, which is a firm indication of an invasive agent. By analogy, the same might apply to the other cell types involved.

Independently, the presence of a retrovirus in the affected target organ could initiate the chain of reactions which cause the induction and perpetuation of autoimmunity. In humans, as has been shown for ovine and caprine lentiviruses [135], these viruses could induce the production of a highly specific viral IFN molecule which may act as a co-factor for the activation of MHC products and other accessory molecule genes in the surrounding cells. The striking picture of the autoimmune pathology is that the micro-environment at the site of the lesion has, at one stage or another, an abundance of various cytokines. Even epithelial cells may become "immunologically" active by producing, for example, IL-6 or IFN-α, thus allowing the cells to communicate better with capillary cells or invading lymphocytes. So, the classical cytokines, regardless from which cells they are secreted, would have a simple paracrine effect, maintaining the "activation" of the affected cells. The practical consequence would be continued boosting of the inappropriate expression of MHC and accessory molecules in situ.

With regard to the putative retrovirus, the ensuing intracellular production and release of the highly specific viral IFN would aim at restraining viral replication without, however, completely preventing it. Thus, viral replication would continue to occur but the number of copies produced would be low and the next result, therefore, limited for the potential hazardous pathogenicity of the virus. However, the process has to be magnified if the pathology is to progress and the disease ultimately to become manifest. Common viruses could thus be implicated at a later stage. They could act as transcription enhancing/promoters for retroviral replication, in a similar way that herpes virus interacts with HIV-1 [136]. It is of interest that CMV sequences have been found in the DNA of PBMC of newly diagnosed type I diabetics [137], and their presence may underline a similar mode of action.

Once class II molecules are expressed on the target epithelial cells, we believe that these cells can indeed present surface autoantigens to autoreactive T cells. This conclusion is based upon the data of the experiments cited above, where class II positive thyrocytes isolated from autoimmune thyroids were able to stimulate specific autoreactive T cell clones expanded from the same glands [48]. Clearly, a particular HLA haplotype (e.g. DR4-DQw8) might display a specific affinity for the surface expressed autoantigen and this would enable it to operate as a more efficient "presenter" than other haplotypes (e.g DR2 or DQw7) (for review see [124]). The example refers to the known positive and negative HLA associations in type I diabetes but the concept could be extended to the genetics of other autoimmune disorders (for review see [138,139]). The first identification of the crystallographic structure of the HLA class I molecule [64] has provided a direct demonstration of how peptides are sited on the "Björkman groove", thus polymorphisms of the HLA molecules may now be attributed distinct functional significances [140]. It is predicted that class II molecules would have similar properties [141] and peptides would then preferentially attach on the surface of the MHC molecules according to the configuration of the groove. Advantageous presentation of certain antigens compared with others can be readily envisaged.

If the same applies to class II molecules on the endocrine cell, efficient presentation of autoantigens would trigger the activation of T cells. After all, the T cell clones expanded from the autoimmune thyroids were of the helper phenotype, and could be regarded as the effector cells responsible for sending signals to autoantigen-specific B lymphocytes, the cells which ultimately produce circulating autoantibodies. It is important to emphasize that B lymphocytes are the least tolerant cells among immunocytes, possessing in normal individuals, for example, immunoglobulin receptors for common autoantigens, such as thyroglobulin [142] and DNA (for review see [143]), thus any positive signal from autoreactive Th cells can readily activate them and make them secrete autoantibodies.

But how are autoreactive T cells generated in the first instance? There is now convincing evidence that some potentially autoreactive T cells are eliminated in the thymus before birth, during the process of T cell maturation and education in the gland [144–147]. Interestingly, the same concept seems now to apply to immature T cells expressing Tγ/δ receptors [148]. On the other hand, other workers have suggested that some T cells escape deletion in the thymus and become anergic, despite maintaining all their potential to recognize and interact with peripheral autoantigens [149]. This is best exemplified in certain experimental models where common autoantigens (e.g. thyroglobulin, basic myelin protein) injected into animals with appropriate immunization protocols can cause disease, such as thyroiditis and encephalomyelitis. The ease with which these autoantigens can break tolerance have strongly supported the view that the autoreactive T cells which recognize them must pre-exist in these animals, as best evidenced by the success to passively transfer disease with T cells to naive syngeneic recipients (for review see [150]).

However, it is important to emphasize that to break tolerance successfully necessitates particular immunization procedures (i.e. the concomitant use of adjuvants), and this may lead to drastic changes in the configuration of the autoantigen molecule, which may thus present differently to T cells, which then recognize it as "foreign" rather than as "self". Interestingly, the animals immunized in this manner produce autoantibodies to several antigenic determinants present in the molecule, while autoimmune patients elicit a very selective humoral response, with the autoantibodies recognizing only a few epitopes [151,152]. This difference is crucial and might argue against the "actual" existence of pre-formed autoreactive T cells in normal individuals. On the other hand, this is not always the case, because we know that tolerance can be broken and disease produced with less harsh immunization protocols, such as lipopolysaccharide (LPS) + antigen [153]. In this situation the injected molecule is not denatured and it should, therefore, maintain its configuration as in the native state. If this is the case, potentially autoreactive T cells must, therefore, be present in the peripheral immune compartments. However, it is important to point out that this procedure is particularly successful in so-called susceptible or high responder animals for that particular antigen [154]. Perhaps this indicates that these animals are different from the so-called low responders in that the former have acquired autoreactive T cells, while in the latter they do not exist at all. Could the same concept apply to humans, accounting for the difference between "genetically" and "non-genetically" susceptible individuals?

Even if we favour the idea of a complete deletion of potentially autoreactive T cells in the thymus, we are prepared, at present, to keep an open mind on this critically important, but still unresolved, question in immunology. Whatever the answer will be, one could again invoke a retrovirus in the generation or expansion of pre-existing autoreactive T cells. If a retroviral transcription occurred in the T cells (so far, retroviruses have shown a predilection for immunocytes), this could have two potentially distinct, but not mutually exclusive, effects. On one hand, the process could lead to amplification of the number of existing autoreactive T cells in an anergic state. On the other hand, retroviruses could also act by re-arranging TCR genes, so that the surface receptor on the T cells develops a new configuration which then binds perfectly to the autoantigen expressed on the target cells, as recent experimental evidence seems to suggest [155].

Autoimmunity is a long chronic pathological process and it takes several years before a crucial number of cells in the affected target organ are destroyed. Even if it is true that the first limb of the immune system has been activated against Th and B lymphocytes leading to the production of autoantibodies, it is also evident that this humoral attack is insufficient to complete the killing by itself. The "vicious circle" must be closed by the Tc cells. Their production is apparently a late event during the process, because if present in sufficient numbers at the time when autoantibodies are first detectable in the genetically predisposed individuals, they should be able to destroy epithelial cells rapidly. This assumption derives from the evidence produced in the pancreatic transplantation experiments in the identical twins [84]. On the other hand retroviruses could act independently and at different time intervals in the two main subsets of T lymphocytes. Should they interact with Tc cells, re-arranging TCR genes, at the pathological climax then β cells or other endocrine cells will not escape from inevitable destruction. In this context, it is important to recall that class I molecules are hyperexpressed on tissues affected by autoimmunity, so offering an even better chance for Tc cells to interact with the surface-Tc expressed autoantigen.

The long latency which precedes the onset of clinical symptoms could also be influenced by other factors. Diminished activity and functional abnormalities of Ts cells have been advocated in a variety of organ-specific autoimmune disorders, not excluding type I diabetes [156]. Despite the controversies raised around the concept of an organ-specific Ts cell defect in endocrine autoimmunity (see a recent debate which followed [156]), on balance it seems plausible and it remains of practical value for the design of future strategies. Within the context of this discussion, either Ts cells or other still unidentified suppressor factors, could be responsible for keeping the gradual development of autoimmune recognition against the target organ under control, at least for a reasonable period of time.

In contrast to the spontaneous animal models of type I diabetes (for review see [157]), in the human disease macrophages have proffered no decisive clues (for review see [158]). If a role has to be attributed to these cells, the only justifiable one is that of clearing the debris of dead cells at the end of the damaging process.

In conclusion, the pathogenesis of autoimmune diseases is still largely unknown. The subject, however, has attracted a great deal of interest among immunologists. It is only by further dissecting the various limbs of the immune response against endocrine or other epithelial cells and. most importantly, by trying to formulate novel hypotheses, sometimes against accepted dogmas [159] that the final picture will be disclosed and the aetiopathogenesis of these diseases established. When this goal is reached, it will enable effective and safe prevention trials to be designed, so that type I diabetes and other related autoimmune disorders will ultimately be prevented.

Acknowledgements. We would like to thank all the colleagues who contributed to the development of this subject throughout the years, and in particular Dr Betty Dean for constructive discussion and help in editing this review. We are also grateful to the British Diabetic Association, the Medical Research Council, The Wellcome Trust Foundation, the Juvenile Diabetes Foundation (JDF) USA, and Novo Research Institute (Copenhagen) for their continuous generous support. A.C. is in receipt of a Fellowship from the JDF; U.D. was supported by a grant from the Deutsche Forschungsgemeinschaft, FRG; and A.M. is supported by a grant from Uehara Memorial Foundation, Tokyo, Japan. Mrs Kathy Cusack is greatly appreciated for patiently typing this manuscript.

References

1. Acuto O, Reinherz EL (1985) The human T cell receptor: structure and function. New Engl J Med. 312:1110–1111
2. Marrack P (1987) New insight into antigen recognition. Science 238:1311–1313
3. Marrack P, Kappler J (1988) The cell repertoire for antigen MHC. Immunol Today 9:308–315
4. Pujol-Borrell R, Todd I (1987) Inappropriate HLA class II in autoimmunity: is it the primary event? In: Doniach D, Bottazzo GF (eds) Endocrine and other organ orientated autoimmune disorders. Bailliere Tindall, London, pp 1–27 (Clinical immunology and allergy, vol 1)
5. Bodmer JG, Marsh SGE, Ekkehard A (1989) Nomenclature for factors of the HLA system. Immunol Today 11:1–33
6. Zinkernagel RM, Doherty RC (1974) Immunological surveillance against altered self-components by sensitized T lymphocytes in lymphocytic choriomeningitis. Nature 251:547–548
7. Battisto JR, Plate J, Shearer G (eds) (1988) In: Cytotoxic T cells. Biology and relevance to disease. Ann NY Acad Sci 532–542
8. Unanue ER, Allen PM (1987) The basis of the immunoregulatory role of macrophages and other accessory cells. Science 235:551–553
9. Bottazzo GF, Todd I, Mirakian R, Belfiore A, Pujol-Borrell R (1986) Organ-specific autoimmunity: a 1986 overview. Immunol Rev 94:137–169
10. Wiman K, Curman B, Forsum V, Klarskog L, Malmas-Tjernlund L, Rask L, Traghardh L, Peterson PA (1978) Occurrence of Ia antigens on tissues of non-lymphoid origin. Nature 276:711–713
11. Forsum U, Klareskog L, Peterson PA (1979) Distribution of Ia antigen-like molecule on non-lymphoid tissues. Scand J Immunol 9:343–349
12. Natali PG, De Martino C, Quaranta V, Nicotra MR, Frezza F, Pelligrino MA, Ferrone S (1981) Expression of Ia like antigens in normal human non-lymphoid organs. Transplantation 31:75–78
13. Nixon DF, Ting JP, Frelinger A (1982) Ia antiens on non-lymphoid tissues: their origins and functions. Immunol Today 3:339–342
14. Bland PW, Warren LG (1986) Antigen presentation by epithelial cells of the rat small intestine. II. Selective induction of suppressor T cells. Immunology 58:9–14
15. Fuggle SV, Errasti P, Daar AS, Fabre JW, Ting A, Morris PJ (1983) Localisation of major

histocompatibility complex (HLA-A,B,C and DR) antigens in 46 kidneys. Transplantation 35:385–390

16. Hall BM, Bishop GA, Duggin GG, Horvath JS, Philips J, Tiller DJ (1984) Increased expression of HLA DR antigens on renal tubular cells in renal transplants: relevance to the rejection response. Lancet ii:247–251
17. Bovo P, Mirakian R, Merigo F, Angelini G, Cavallini G, Rizzini P, Bottazzo GF, Scuro LA (1987) HLA molecule expression on chronic pancreatitis specimens: is there a role for autoimmunity? A preliminary study. Pancreas 2:350–356
18. Baekkeshov S, Kanatsuna T, Klareskog L, Nielsen DA, Peterson PA, Rubistein AH, Steiner DF, Lernmark A (1981) Expression of major histocompatibility antigens on pancreatic islet cells. Proc Natl Acad Sci USA 78:6456–6460
19. McNicol AM (1986) Class II MHC antigen expression in adrenal cortex. Lancet ii:1282–1283
20. Khoury EL, Greenspan JS, Greenspan FS (1987) Adrenocortical cells of the zona reticularis normally express HLA DR antigenic determinants. Am J Pathol 127:580–591
21. Pujol-Borrell R, Todd I, Londei M, Foulis A, Feldmann M, Bottazzo GF (1986) Inappropriate MHC Class II expression by thyroid follicular cells in thyroid autoimmune disease and by pancreatic beta cells in Type I diabetes. J Molec Biol Med 6:159–165
22. Bottazzo GF, Dean BM, McNally JM, Hackay EH, Swift PGF, Gamble DR (1985) In situ characterisation of autoimmune phenomena and expression of HLA molecules in the pancreas in diabetic insulitis N Engl J Med 313:353–360
23. Caforio ALP, Stewart JT, Bonifacio E, Burke M, Davies MJ, McKenna WJ, Bottazzo GF (1990) Inappropriate Major Histocompatibility complex expression on cardiac tissue in dilated cardiomiopathy. Relevance for autoimmunity. J Autoimmunity 3:187–200
24. Pujol-Borrell R, Hanafusa T, Chiovato L, Bottazzo GF (1983) Lectin-induced expression of DR antigen on human cultured follicular thyroid cells. Nature 305:71–73
25. Hanafusa T, Pujol-Borrell R, Chiovato L, Russell RCG, Doniach D, Bottazzo GF (1983) Aberrant expression of HLA-DR antigen on thyrocytes in Graves' disease: relevance for autoimmunity. Lancet ii:1111–1115
26. Lucas-Martin A, Foz-Sala M, Todd I, Bottazzo GF, Borrell R (1988) Occurrence of thyrocyte Class II expression in a wide variety of thyroid diseases: relationship with lymphocyte infiltration and thyroid autoantibodies. J Clin Endocrinol Metab 66:367–375
27. Aichinger G, Fill H, Wick G (1985) In situ immune complexes: lymphocyte subpopulations and HLA-DR positive epithelial cells in Hashimoto thyroiditis. Lab Invest 52:132–140
28. Jansson R, Karlsson A, Forsum U (1984) Intrathyroidal HLA-DR expression and T lymphocyte phenotypes in Graves' thyrotoxicosis, Hashimoto's thyroiditis and nodular colloid goiter. Clin Exp Immunol 58:264–272
29. Grubeck-Loebenstein B, Derfler K, Kassal H, Knapp W, Krisch K, Liszka K, Smyth PPA, Waldhausl W (1985) Immunological features of non-immunogenic hyperthyroidism. J Clin Endocrinol Metab 60:150–155
30. Grubeck-Loebenstein B, Kassal H, Smyth RPA, Krisch K, Waldhausl W (1986) The prevalence of immunological abnormalities in endemic simple goitre. Acta Endocrinol 113:508–512
31. Todd I, Pujol-Borrel R, Abdul-Karim BAS, Hammond LJ, Feldmann M, Bottazzo GF (1987) HLA-D subregion expression by thyroid epithelium in autoimmune thyroid diseases and induced in vitro. Clin Exp Immunol 69:532–538
32. Czarnocka B, Ruf J, Ferrand M, Carayon P, Lissitzky S (1985) Purification of the human thyroid peroxidase and its identification as the microsomal antigen involved in autoimmune thyroid diseases. FEBS Lett 190:147–152
33. Portman L, Hamada N, Heinrich G, DeGroot LJ (1985) Anti-thyroid peroxidase antibody in patients with autoimmune thyroid disease: possible identity with anti-microsomal antibody. J Clin Endocrinol Metab 61:1001–1003
34. Kotani T, Umeki K, Matsunaga S, Kata E, Ohtaki S (1986) Detection of autoantibodies to thyroid peroxidase in autoimmune thyroid disease by micro ELISA and immunoblotting. J Clin Endocrinol Metab 62:928–934
35. Libert F, Ruel J, Ludgate M, Swillens S, Alexander N, Vassart G, Dinsart C (1987) Thyroperoxidase, an auto-antigen with a mosaic structure made of nuclear and mitochondrial gene modules. EMBO J 6:4193–4196
36. Todd I, Abdul-Karim BAS, Pujol-Borrell R, Feldmann M, Bottazzo GF (1986) Dissection and characterisation of the spontaneous and induced expression of HLA-D/DR by thyroid epithelium. In: Medeiros-Nero G, Gaitan E (eds) 'Frontiers in thyroidology', vol 2. Plenum Press, New York, pp 1587–1592

37. Lindhal G, Hedfors E, Klareskog L, Forsum U (1985) Epithelial HLA-DR expression and T lymphocytes subsets in salivary glands in Sjögren's syndrome. Clin Exp Immunol 61:475–482
38. Harley JB, Reichlin M, Arnett FC, Alexander EL, Bias WB, Provost TT (1986) Gene interaction at HLA-DQ enhances autoantibody production in primary Sjögren's syndrome. Science 232:1145–1147
39. Foulis AK, Farquharson MA (1986) Aberrant expression of HLA-DR antigens by insulin containing beta cells in recent onset type 1 (insulin-dependent) diabetes mellitus. Diabetes 35:1215–1224
40. Foulis AK, Farquharson A, Hardman R (1987) Aberrant expression of Class II major histocompatibility complex molecules by B cells and hyperexpression of Class I major histocompatibility complex molecules by insulin containing islets in Type I (insulin-dependent) diabetes mellitus. Diabetologia 30:333–343
41. Bottazzo GF, Todd I, Pujoll-Borrel R, Londe M, Feldman M (1986) New ideas on the pathogenesis of autoimmune disease. In: Triger DR (ed) Advanced Medicine, vol 22. Transmedica Europe Ltd, UK pp 66–76
42. Mirakian R, Hill S, Richardson A, Milla PJ, Walker-Smith JA, Bottazzo GF (1988) HLA product expression and lymphocyte subpopulations in jejunum biopsies of children with idiopathic protracted diarrhoea and enterocyte autoantibodies. J Autoimmunity 1:263–277
43. Ballardini G, Bianchi FB, Mirakian R, Fallani M, Pisi E, Bottazzo GF (1987) HLA-A,B,C, HLA-D/DR and HLA-D/ DQ expression on unfixed liver biopsy reactions from patients with chronic liver disease. Clin Exp Immunol 70:35–46
44. Kallemberg CGM, Schilizzi BM, Beaumont F, Popperwe S, De Leij L, The TH (1987) Expression of class II MHC antigens on alveolar epithelium in filmosing alveolitis. Clin Exp Immunol 67:182–190
45. Bottazzo GF, Pujol-Borrell R, Hanafusa T, Feldman M (1983) Role of aberrant HLA-DR expression and antigen presentation in induction of endocrine autoimmunity. Lancet ii:1115–1119
46. Davies TF (1985) Cocultures of human thyroid monolayer cells and autologous T cells: impact of HLA Class II antigen expression. J Clin Endocrinol Metab 61:418–422
47. Londei M, Lamb JR, Bottazzo GF, Feldmann M (1984) Epithelial cells expressing aberrant MHC Class II determinants can present antigen to cloned human T cells. Nature 312:639–641
48. Londei M, Bottazzo GF, Feldmann M (1985) Human T cell clones from autoimmune thyroid glands: specific recognition of autologous thyroid cells. Science 228:85–89
49. Weetman AP, Volkman DJ, Burman KD, Gerrard TL, Fauci AS (1985) The in vitro regulation of human thyrocyte HLA-DR antigen expression. J Clin Endocrinol Metab 61:817–824
50. Eguchi K, Otsubo T, Kawabe Y, Ueki Y, Fukuda T, Matsunaga M, Shimomura C, Ishikawa N, Tezuka H, Nakao H, Ito K, Morimoto C, Nagataki S (1987) The remarkable proliferation of helper T cells subsets in response to autologous thyrocytes and intrathyroidal T cells from patients with Graves' disease. Clin Exp Immunol 70:403–410
51. Eguchi K, Otsubo T, Kawabe Y, Shimomura C, Ueki Y, Nakao H, Tezuka H, Matsunaga M, Fukuda T, Ishikawa N, Ito K, Nagataki S (1988) Synergy in antigen presentation by thyroid epithelial cells and monocytes from patients with Graves' disease. Clin Exp Immunol 72:84–87
52. Grubeck-Loebenstein B, Londei M, Greenall C, Pirich K, Kassal H, Waldhäusl W, Feldman M (1988) Pathogenic relevance of HLA-Class II expressing thyroid follicular cells in non toxic goitre and in Graves' disease. J Clin Invest 81:1608–1612
53. Stoolman LM (1989) Adhesion molecules controlling lymphocyte migration. Cell 56:907–910
54. Makgoba MW, Sanders ME, Shaw S (1989) The CD_2-LFA_3 and LFA-1-ICAM-1 pathways: relevance to T cell recognition. Immunol Today 10:417–422
55. Altmann DM, Hogg N, Trowsdale J, Wilkinson D (1989) Cotransfection of ICAM-1 and HLA-DR reconstitutes human antigen-presenting cell function in mouse L cells. Nature 338:512
56. Campbell IL, Cutri A, Wilkinson D, Boyd AW, Harrison LC (1989) Intercellular adhesion molecule 1 is induced on isolated endocrine islet cells by cytokines but not by retrovirus infection. Proc Natl Acad Sci USA 86:4282–4286
57. Pujol-Borrell R, Vives M, Badenas J, Foz M, Buscema M, Soldevilla G, Bottazzo GF (1989) Modulation of intercellular adhesion molecule-1 (ICAM-1) in human islet cells. Diabetes 38(Suppl 22): 733A
58. Mirakian R, Richardson A, Milla PJ, Walker-Smith JA, Unsworth J, Savage MO, Bottazzo GF (1986) Protracted diarrhoea of infancy: evidence in support of an autoimmune variant. Br Med J 293:1132–1136
59. Hill S, Casaseca T, Milla P, Bottazzo GF, Mirakian R (1989) Adhesion molecule expression in

autoimmune enteropathy. In: Proceedings of the 5th international congress of gut immunology, London, July 1989 (in press)
60. Allison J, Campbell IL, Morahan G, Mandel TE, Harrison LC, Miller JFAP (1988) Diabetes in transgenic mice resulting from over-expression of Class I histocompatibility molecules in pancreatic β cells. Nature 333:529–533
61. Lo D, Burkly LC, Widera G, Cowing C, Flavell RA, Palmiter RD, Brinster RL (1988) Diabetes and tolerance in transgenic mice expressing Class II MHC molecules in pancreatic beta cells. Cell 53:159–163
62. Sarvetnick N, Liggit D, Pitts SL, Hansen SE, Stewart TA (1988) Insulin-dependent diabetes mellitus induced in transgenic mice by ectopic expression of Class II MHC and interferon-gamma. Cell 52:773–782
63. Parham P (1988) Intolerable secretion in tolerant transgenic mice. Nature 333:500–503
64. Björkman PJ, Saper MA, Samraoui B, Bennett WS, Strominger JL, Wiley DC (1987) Structure of the human Class I histocompatibility antigen HLA-DR2. Nature 329:506–509
65. Bohme J, Haskins K, Stecha P, van Ewijk W, Lemeur M, Gerlinger P, Benoist C, Mathis D (1989) Transgenic mice with I-A on islet cells are normoglycemic but immunologically intolerant. Science 244:1179–1183
66. Stockinger B, Pessara U, Lin RH, Habich J, Grez M, Koch N (1989) A role of Ia-associated invariant chains in antigen processing and presentation. Cell 56:683–689
67. Pujoll-Borrell R, Bottazzo GF (1988) Puzzling diabetic transgenic mice: do they contain a lesson for human Type I diabetes? Immunol Today 9:303–306
68. Markmann J, Lo D, Naji A, Palmiter RD, Brinster RL, Heber-Katz E (1988) Antigen presenting function of Class II MHC expressing beta cells. Nature 336:476–479
69. Balkwill FR, Burke F (1989) The cytokine network. Immunol Today 10:299–304
70. Todd I, Pujol-Borrell R, Hammond LJ, Bottazzo GF, Feldmann M (1985) Interferon-gamma induces HLA-DR expression by thyroid epithelium. Clin Exp Immunol 61:265–273
71. Basham TY, Nickoloff BJ, Merigan TC, Morhenn UB (1984) Recombinant gamma interferon induces HLA-DR expression on cultured human keratinocytes. J Invest Dermatol 83:88–92
72. Houghton A, Thomson TM, Gross D, Oettgen HF, Old LJ (1984) Surface antigens of melanoma and melanocytes: specificity of induction of Ia antigens by human interferon-gamma. J Exp Med 160:255–269.
73. Lu CY, Changelian PS, Unanue ER (1984) Alpha-fetoprotein inhibits macrophage expression of Ia antigens. J Immunol 132:1722–1727
74. De Preval C, Lisowska-Grospierre B, Loche M, Griscelli C, Mach B (1985) A trans-acting class II regulatory gene unlinked to the MHC controls expression of HLA Class II genes. Nature 308:291
75. Flavel RA, Allen H, Burkly, Sherman DH, Werneck GL, Widers G (1986) Molecular biology of the MHC histocompatibility complex. Science 233:437–443
76. Todd I, Pujol-Borrell R, Hammond LJ, McNally JM, Feldmann M, Bottazzo GF (1987) Enhancement of thyrocyte Class II expression by thyroid stimulating hormone. Clin Exp Immunol 69:524–531
77. Wenzel BE, Gutekunst R, Mansky T, Schultek T, Scriba PC (1986) Thyrotropin and IgG from patients with Graves' disease induce class-II antigen on human thyroid cells. In: Drexhage HA, Wiersinga WM (eds) The thyroid and autoimmunity. Excerpta Medica, Amsterdam, pp 141–144
78. Buscema M, Todd I, Deuss U, Hammond LJ, Pujol-Borrell R, Mirakian R, Bottazzo GF (1989) Influence of TNF-alpha on the modulation by IFN-gamma of HLA Class II in human thyroid cells and its effect on IFN-gamma binding. J Clin Endocrinol Metab 69:433–439
79. Todd I, Hammond LJ, Feldmann M, Bottazzo GF Epidermal growth factor and transforming growth factor-alpha suppress HLA Class II induction in human thyroid epithilial cells. Immunology (in press)
80. Guerin V, Todd I, Hammond LJ, Bottazzo GF (1990) Suppression of HLA Class II expression by interferon-α1 Clin Exp Immunol 79:341–345
81. Pujol-Borrell R, Todd I, Doshi M, Gray D, Feldmann M, Bottazzo GF (1986) Differential expression and regulation of MHC products in the endocrine and exocrine cells of the human pancreas. Clin Exp Immunol 65:128–139
82. Shalaby MR, Aggarwal BB, Rinderknecht E (1985) Activation of human polymorphonuclear neutrophil functions by interferon-gamma and tumor necrosis factor. J Immunol 135:2069–2073
83. Pujol-Borrell R, Todd I, Doshi M, Bottazzo GF, Sutton R, Gray D, Adolf G, Feldmann M (1987) HLA Class II induction in human islet cells by interferon-gamma plus tumour necrosis factor or lymphotoxin. Nature 326:304–306

84. Sibley RK, Sutherland DER (1987) Pancreas transplantation: an immunohistologic and histopathologic examination of 100 grafts. Am J Pathol 128:151–170
85. Mauheroff T, Belfiore A, Pujol-Borrell R, Bottazzo GF (1990) Growth inhibition on human endothelial cells by human recombinant tumour necrosis factor α and interferon γ. Eur J Cell Biol (in press)
86. Leiborich SJ, Polverini PJ, Shepard HM, Wiseman DM, Shively V, Nuseir N (1987) Macrophage-induced angiogenesis is mediated by TNFα. Nature 329:630–637
87. Frater-Schroder M, Risaw W, Hallmann R, Grantschi P, Bohlen P (1987) TNFα, important inhibition of endothelial cell growth in vitro is angiogenic in vivo. Proc Natl Acad Sci USA 84:5277–5282
88. Ruddle NH (1987) TNFα and related cytokines. Immunol Today 8:129–130
89. Deuss U, Buscema M, Bottazzo GF (1988) Rehabilitation of cytokines in thyroid autoimmunity: are they only exerting negative effects? In: Nagataki S, Torizuka K (eds) The thyroid. Excerpta Medica, Amsterdam pp 44–50 (International congress series, no. 796)
90. Beutler B, Cerami A (1988) The common mediator of shock, cachexia, and tumor necrosis. Adv Immunol 42:213–231
91. Kroemer G, Brezinshik HP, Faessler R, Schaventein K, Wick G (1988) Physiology and pathology of an immunoendocrine loop. Immunol Today 9:163–165
92. Harrison LC, Campbell IL (1988) Cytokines: an expanding network of neuro-inflammatory hormones. Molec Endocrinol 2:1151–1156
93. Calkin JH, Siegel MM, Nankin HR, Lim T (1988) Interleukin-1 inhibits Leydig cell steroidogenesis in primary culture. Endocrinology 123:1605–1610
94. Verhoeven G, Cailleaw J, Van Damme J, Biliau A (1988) Interleukin-1 stimulates steroidogenesis in cultured rat Leydig cell. Molec Cell Endocrinol 57:51–60
95. Burman P, Totterman TH, Oberg K, Karlsson FA (1986) Thyroid autoimmunity in patients on long term therapy with leukocyte derived interferon. J Clin Endocrinol Metab 63:1086–1090
96. Fentiman IS, Balkwill FR, Thomas BS, Todd I, Bottazzo GF (1988) An autoimmune aetiology for hypothyroidism following interferon therapy for breast cancer. Eur J Cancer Clin Oncol 24:1299–1303
97. Atkins MB, Mier JW, Parkinson DR, Gould JA, Berkman EM, Kaplan MM (1988) Hypothyroidism after treatment with IL-2 and LAK cells. N Engl J Med 1988. 318:1557–1561
98. Karlsson FA, Burman P, Oberg , Karisson-Pane A (1989) Thyroid autoimmunity induced by interferon. In: Symposium on thyroiditis. Homberg, FRG. October 1989 (in press)
99. Bendtzen K (1989) Immune hormones (cytokines); pathogenic role in autoimmune rheumatic and endocrine diseases. Autoimmunity 2:177–189
100. Mandrup-Poulsen T, Helquist S, Molvig J, Wogensen LD, Nerup J (1989) Cytokines as immune effector molecules in autoimmune endocrine diseases with special references to insulin-dependent diabetes mellitus. Autoimmunity 4:191–218
101. Campbell IL, Iscaro A, Harrison LC (1988) IFNγ, and TNFα: cytotoxity to endocrine islets of Langerhans. J Immunol 141:2325–2329
102. Pukel C, Baqueriro H, Rabinovitch (1988) Destruction of rat islet cell monolayers by cytokines. Synergistic interactions of IFN-γ, TNFα, lymphotoxin and IL-1. Diabetes 37:133–136
103. Foulis AK, Farquharson MA, Meager A (1987) Immunoreactive α-interferon in insulin-secreting B-cells in Type I diabetes mellitus. Lancet ii:1423–1437
104. Campbell IL, Cutri A, Wilson A, Harrison LG (1989) Evidence for IL-6 production by and effects on the pancreatic β cells. J Immunol 143:1188–1191
105. Foulis AK (1989) In Type I diabetes, does a non-cytopathic viral infection of insulin-secreting B-cells initiate the disease process leading to their autoimmune destruction? Diab Med 6:666–674
106. Massa PT, Dörries R, Meulen V (1986) Viral particles induce Ia antigen expression on astrocytes. Nature 320:543–546
107. Albino AP, Houghton AN, Eisingor M (1986) Class II histocompatability antigen expression in human melanocytes transformed by Harvey murine sarcoma virus (Ha-MSV) and Kirsten MSV retrovirus. J Exp Med 164:1710–1712
108. Suciu-Foca N, Rubinstein P, Popovic M, Gallo RC, King DW (1984) Reactivity of HTLV transformed human T-cell lines to MHC Class II antigens. Nature 312:275–277
109. Majello B, La Mantia G, Simeone A, Boncinelli E, Lania L (1985) Activation of major histocompatibility complex Class I m RNA containing an Alu-like repeat in polyoma virus-transformed rat cells. Nature 314:457–459
110. Brickell PM, Latchman DS, Murphy D, Willison K, Rigby PWJ (1985) The Class I major histocompatability antigen gene activated in a line of SV40 transformed mouse cells is H-2D^{d} not Qa/Tla. Nature 316:162–163

111. Anderson M, Pääbo S, Nilsson T, Peterson P (1985) Impaired intracellular transport of Class I MHC antigens as a possible means for adenovirus to evade immune surveillance. Cell 43:215–222
112. Belfiore A, Mauerhoff T, Pujol-Borrell R, Mirakian R, Bottazzo GF (1987) Constitutive expression of HLA Class II molecules in human thyroid cells transfected with SV-40. In: Pinchera A, Ingbar SH, McKenzie JM, Fenzi GF (eds) Thyroid autoimmunity. Plenum Press, New York pp 465–467
113. Rayfield JR, Ishimura K (1987) Environmental factors and insulin-dependent diabetes mellitus. Diab Metab Rev 3:925–957
114. Karam JH, Lewitt PA, Young CW, Nowlain RE, Frankel BJ, Fujixa H, Freedman ZR (1980) Insulopenic diabetes after rodenticide (Vacor) ingestion. A unique model of acquired diabetes in man. Diabetes 29:971–978
115. Helgason T, Jonasson MR (1981) Evidence for a food additive as a cause of ketosis-prone diabetes. Lancet ii:716–720
116. Leiter EH (1990) The role of environmental factors in modulating insulin dependent diabetes. In: This volume, pp 39–51
117. Bottazzo GF, Pujol-Borrell R, Gale EAM (1986) Autoimmunity and diabetes: progress, consolidation and controversy. In: Alberti KGM, Krall LP (eds) The diabetes annual, vol 2. Amsterdam, pp 13–29
118. Gorsuch AN, Spencer KM, Lister J, McNally JM, Dean BM, Bottazzo GF, Cudworth AG (1981) Evidence for a long prediabetic period in type I (insulin-dependent) diabetes mellitus. Lancet ii:1363–1365
119. Bonifacio E, Bingley P, Shattock M, Dean BM, Dunger D, Gale EAM, Bottazzo GF (1989) Quantification of islet-cell antibodies and prediction of insulin-dependent diabetes. Lancet 335:147–149
120. Johnston C, Millward BA, Hoskins P, Leslie RDG, Bottazzo GF, Pyke DA (1984) Islet-cell antibodies as predictors of the later development of type 1 (insulin-dependent) diabetes. A study in identical twins. Diabetologia 32:382–386
121. Srikanta S, Ganda OP, Jackson RA, Gleason RE, Kaldany A, Garovoy MR, Milford EL, Carpenter GB, Soeldner JS, Eisenbarth GS (1983) Type 1 diabetes mellitus in monozygotic twins: chronic progressive beta cell disfunction. Ann Intern Med 99:320–326
122. Betterle C, Presotto F, Pedini B, Moro L, Slack RS, Zanette F, Zanchetta R (1987) Islet cell and insulin autoantibodies in organ specific autoimmune patients. Their behaviour and predictive value for the development of Type 1 (insulin-dependent) diabetes mellitus. A 10 year follow-up study. Diabetologia 30:292–297
123. Bruining GJ, Molenaar JL, Grobee DE, Hofman A, Scheffer GJ, Brüining HA, Debrugn AM, Valkenburg NA (1989) Ten year follow up study of islet cell antibodies and childhood diabetes mellitus. Lancet i:1100–1103
124. Bottazzo GF (1983) Death of a beta cell: homicide or suicide? Diab Med 3:119–130
125. Ziamiecki A, Kronner G, Mueller RG, Hala K, Wick G (1988) Iv22, a new endogenous avian leukosis virus locus found in chickens with spontaneous autoimmune thyroiditis. Arch Virol 100:267–270
126. Leiter EH, Wilson GL (1988) Viral interactions with pancreatic B cells. In: Lefebvre PJ, Pipeleers DH (eds) The pathology of the endocrine pancreas in diabetes. Springer, Berlin, pp 86–105
127. Ciampolillo A, Marini V, Mirakian R, Buscema M, Schulz T, Pujol-Borrell R, Bottazzo GF (1989) Retrovirus-like sequences in Graves' disease: implications for human autoimmunity. Lancet i:1096–1100
128. Ciampolillo A, Marini V, Mirakian R, Pujol-Borrell R, Bottazzo GF (1989) Progress towards the identification of novel retroviruses in thyroid autoimmunity. In: Symposium on thyroiditis. Homberg, FRG, October 1989 (in press)
129. Streeter PR, Berg EL, Rouse BTN, Bargatze RL, Butcher EC (1988) A tissue-specific endothelial cell molecule involved in lymphocyte homing. Nature 331:41–46
130. Jalkanen S, Steere AC, Fox RI, Butcher EC (1986) A distinct endothelial cell "recognition" system that controls lymphocyte traffic into inflamed synovium. Science 233:556–560
131. Pober JS, Collins T, Grimbone MA, Libby P, Reiss CS (1986) Inducible expression of class II major histocompatibility complex antigens and the immunogenicity of the vascular endothelium. Transplantation 41:141–146
132. Rafferzeder M, Bechtner G, Greil W, Gärtner R (1988) Isolated thyroid follicles secrete a paracrine, heparin-binding growth factor for endothelial cells. Acta Endocrinol 117:58–62
133. Foulis AK, Bottazzo GF (1988) Insulitis in the human pancreas. In: Lefebvre PJ, Pipeleers DH (eds) The pathology of the endocrine pancreas in diabetes. Springer, Berlin, pp 41–52

134. Miyazaki A, Hanafusa T, Itoh N, Miyagawa J, Kono N, Tarul S, Kiyotaki C, Yoshizaki K (1989) Demonstration of interleukin 1β on perifollicular endothelial cells in the thyroid glands of patients with Graves' disease. J Clin Endocrinol Metab 69:738–744
135. Kennedy PGE, Narayan O, Ghotbi Z, Hopkins J, Gendelman HE, Clements JE (1985) Persistent expression of Ia antigen and viral genome in Visna-Maedi virus induced inflammatory cells. Possible role of lentivirus induced interferon. J Exp Med 162:1970–1980
136. Mosca JD, Bednarik DP, Raj NBK, Rosen CA, Sodwski JG, Haseltine WA, Pitha PM (1987) Herpes simplex virus Type I can reactivate the transcription of latent HIV. Nature 325:67–70
137. Pak CY, Eun HM, McArthur RG, Yoon JW (1988) Association of cytomegalovirus infection with autoimmune Type 1 diabetes. Lancet i:1–4
138. Todd JA, Acha-Orbea H, Bell JI, Chao N, Frorek Z, Jacob CO, McDermott M, Sinha AA, Timmermann L, Steinman L, McDevitt HO (1988) A molecular basis for MHC Class II-associates autoimmunity. Science 240:1003–1009
139. Trucco M, Dorman JS (1989) Immunogenetics of insulin-dependent diabetes mellitus in humans. In: Critical Review of Immunology, vol 9. CRC Press, Boca Raton, Florida, pp 201–245
140. Garrett TPJ, Saper MA, Bjorkman PJ, Strominger JL, Wiley DC (1989) Specificity pockets for the side chains of peptide antigens in HLA-Aw68. Nature 342:692–696
141. Brown J, Jardetzky T, Saper M, Samraoni B, Bjorkman P, Wiley D (1988) A hypothetical model of the foreign antigen binding site of Class II hystocompatibility molecules. Nature 332:845–852
142. Bankhurst AD, Torrigiani G, Allison AC (1973) Lymphocytes binding human thyroglobulin in healthy people and its relevance to tolerance for autoantigens. Lancet i:226–230
143. Holmberg D, Coutinho A (1985) Natural antibodies and autoimmunity. Immunol Today 6:356–357
144. Kappler JW, Roehm N, Marrak P (1987) T cell tolerance by clonal elimination in the thymus. Cell 49:273–280
145. Kappler JW, Staerz U, White J, Marrack P (1988) Self tolerance eliminates T cells specific for Mls-modified product of the major histocompatibility complex. Nature 332:35–39
146. MacDonald RH, Schneider R, Lees RK, Howe CR, Orbea HA, Festenstein H, Zinkernagel RM, Hengartner H (1988) T-cell receptor V use predicts reactivity and tolerance to Mls-encoded antigens. Nature 332:35–39
147. Bill J, Appel VB (1988) An analysis of T-cell receptor variable region gene expression in major histocompatibility complex disparate mice. Proc Natl Acad Sci USA 85:9184–9188
148. Dent AL, Matis LA, Houshmand F, Widacki SM, Bluestone JA, Hedrick SM (1990) Self-reactive γδT cells are eliminated in the thymus. Nature 343:714–718
149. Rammensee HG, Kroschewski R, Frangoulis B (1989) Clonal anergy induced in mature VB6+ T lymphocytes on immunizing Mls-1^b/Mls-1^a mice with Xls-1^a expressing cells. Nature 339:541–544
150. Cohen IR (1988) T lymphocyte vaccination against autoimmune diseases. In: Cohen IR (ed) Perspectives on autoimmunity. CRC Press, Boca Raton, Florida, pp 21–217
151. Nye L, Ponte De Carvalho LC, Roitt IM (1980) Restriction in the response to autologous thyroglobulin in the human. Clin Exp Immunol 41:252–263
152. Ruf J, Toubert ME, Czarnocka B, Durand-Gorde JM, Ferrand M, Carayon P (1989) Relationship between immunological structure and biochemical properties of human thyroid peroxidase. Endocrinology 125:1211–1218
153. Esquivel PS, Rose NR, Kong YM (1977) Induction of autoimmunity in good and poor responder mice with mouse thyroglobulin and lipopolysaccharide. J Exp Med 145:1250–1263
154. Esquivel PS, Kong YM, Rose NR (1978) Evidence for thyroglobulin-reactive T cells in good responder mice. Cell Immunol 37:14–19
155. Krieg A, Gause WC, Gourley MF, Steinberg AD (1989) A role of endogeneous retroviral sequences in the regulation of lymphocyte activation. J Immunol 143:2448–2457
156. Volpe R (1988) The immunoregulatory disturbance in autoimmune thyroid disease. Autoimmunity 2:55–72
157. Kolb-Bachofen V, Kolb H (1989) A role for macrophage in the pathogenesis of Type 1 diabetes. Autoimmunity 3:145–155
158. Bottazzo GF, Bonifacio E, Al-Sakkaf L, Dean BM, Schwarz G, Mirakian R (1990) Immunogenetic and environmental factors in the pathogenesis of Type 1 (insulin-dependent) diabetes mellitus. An update. In: Samois E (ed) The endocrine pancreas, Raven Press, New York
159. Bottazzo GF, Todd I, Belfiore A, Mirakian R, Pujol-Borrell R (1987) Thyroid and related autoimmune disorders: challenging the dogmas. In Pinchera A, Ingbar SH, McKenzie JM, Fenzi CF (eds) Thyroid autoimmunity. Plenum Press, New York, pp 159–173

Chapter 12

Molecular Mimicry in Autoimmune Disease

T. Dyrberg

Introduction

The basis for a functional immune system is the ability to discriminate between self and non-self. To this end the immune system features a set of unique recognition molecules, the major histocompatibility complex (MHC) molecules of antigen presenting cells, the immunoglobulins of B lymphocytes and the T cell receptors of T lymphocytes. Within this system, a nearly unlimited number of foreign antigens can be recognized, even structures to which the organism has not previously been exposed, yet the organism's own molecules are not recognized. The selective lack of recognition of self antigens is believed to evolve during the maturation of T cells in the thymus [1].

If tolerance to self antigen is broken, the result may be autoimmunity and disease, depending on the target antigen (e.g. the acetylcholine receptor in myasthenia gravis [2]) or the magnitude of the autoimmune response (e.g. immune complex nephropathy [3] in lupus erythomathosus). Although the sequence of events through which the immune system matures is beginning to be unravelled, and the structure and function of the immune recognition molecules are being elucidated [4,5], little is known about how autoimmunity is initiated and which factors determine its progression to disease. Genetic predisposition is known to be important. Genes within the MHC have been found associated with most autoimmune diseases [6], but genes outside the MHC also influence susceptibility to autoimmunity. From studies on type 1 diabetes in monozygotic twins and HLA identical siblings, it has been speculated that MHC genes contribute only 25%–50% of the genetic susceptibility [7]. However, since the concordance rate of many autoimmune diseases in monozygotic twins is below 100%, e.g. 30%–60% in type 1 diabetes [8], it is apparent that factors other than genetics must influence the development of autoimmunity. Of environmental factors, viruses in particular have been implicated [9]. Viruses could trigger autoimmunity through a multitude of mechanisms:

1. Through polyclonal activation of B and T lymphocytes.
2. Through expression of new antigens on cells following virus infection or through alteration of existing antigens, causing them to be recognized by the immune system as foreign molecules.
3. Virus mediated lysis of cells can release cellular antigens to which the organism is not tolerant.
4. Virus infection of lymphocytes or macrophages can result in functional inactivation or lysis of unique subsets of these immune cells.
5. Virus infection can cause an increased expression of MHC class I or II antigens, either directly or indirectly through virus mediated release of lymphokines.
6. Through antigenic mimicry between virus and host cell proteins.

A determinant on a virus protein, or a protein from another microbe may antigenetically mimic a cell protein, a phenomenon that has been termed molecular mimicry [10]. The hypothesis for how this could result in autoimmunity is that an immune response directed against a mimicking epitope on a microbial protein cross-reacts with a cell protein carrying the homologous determinant, and depending on the function of the cellular protein and the magnitude of the cross-reacting immune response, autoimmune disease may follow. Implicit to this hypothesis is that:

1. The immune response which was initiated against the viral protein can be propagated by the cross-reacting cellular protein, even at a time when the virus has been cleared from the organism. Analysis for a viral aetiology at the time of onset may be very difficult, since the clinical onset of an autoimmune disease can be preceded by a long, asymptomatic period, e.g. manifest hyperglycaemia in type 1 diabetes is preceded by several years characterized by islet cell autoimmunity and diminishing beta cell function [11].

2. In order for immunological cross-reactivity to take place, the mimicking viral epitope needs to be sufficiently similar to allow cross-recognition and yet different enough to break tolerance. The molecular basis for this is not known, but genetically determined immune responsiveness is likely to be of importance. The antigenic determinant itself can also play a role. Thus it has recently been shown that single amino acid substitutions in homologous epitopes can influence cross-reactive immune recognition significantly [12].

Molecular Mimicry

Experimental Models

Examples of antibody cross-reactivity between microbial and cellular proteins are plentiful [13]. With the advance of hybridoma technology, it was established that the same antibody could bind to homologous determinants on, for example, a virus and a cell protein [14], and the cross-reactive epitopes could now be more precisely defined [15,16]. Subsequent studies have shown that viruses possess a significant number of mimicking epitopes. Thus, 1%–2% of monoclonal antibodies isolated from mice infected with reo-virus reacted with both the virus and with normal cellular components [17]. Analysis of a large number of monoclonal antibodies to a variety of different human viruses showed that more than 3% bound not only to the virus, but also to normal tissue [18]. Antigenic mimicry between virus and cell proteins is therefore a common occurrence when studied with antibodies selected in vitro, and it is not likely that the mere presence of cross-reactive immune reactants is harmful in itself.

Through widespread use of molecular cloning techniques, sequence information on viral and cellular proteins has become more readily available. This makes it possible to search directly for shared, linear antigenic epitopes between an autoantigen and viral proteins, and the immunological cross-reactivity can then be tested experimentally [13,19]. Following this approach, Fujinami and Oldstone identified a region of hepatitis B virus polymerase that shared 6 out of 10 amino acids with a sequence of myelin basic protein that causes experimental allergic encephalitis in rabbits [20]. Rabbits immunized with the homologous virus sequence evidenced antibodies and peripheral blood mononuclear cells that reacted with intact myelin basic protein. Thus, this experiment directly demonstrated that a virus peptide can elicit an immune response that cross-reacts with an autoantigen. Histological analysis of central nerve system tissue of immunized rabbits showed lesions resembling those found in experimental allergic encephalitis, providing evidence, albeit in an experimental model, to support the hypothesis that autoimmune disease can be caused by molecular mimicry. Other virus proteins have also been reported to display sequence homologies to proteins related to encephalomyelitis and neuritis [21,22].

Mice infected with Theiler's murine encephalomyelitis virus (TMEV) provide another model for human demyelinating diseases. Intracerebral injection of TMEV can induce a chronic demyelinating disease, with histological features similar to those seen in experimental allergic encephalitis [23]. The disease has been suggested to be immunologically mediated, although the contribution of direct viral pathology has not been clarified. A monoclonal antibody to the vp-1 protein of TMEV was shown to cross-react with galactocerebroside, a major component of myelin [24]. Injection of the monoclonal antibody directly into the sciatic nerves of rats was followed by Schwann cell destruction and demyelination, indicating that the antibody bound to myelin structures [25]. These data suggest that more complex antigen cross-reactivities may be involved in demyelinating diseases.

Another extensively studied experimental model, adjuvant arthritis in rats, relates to T cell cross-reactivity [26]. Susceptible rats, immunized intracutaneously with heat-killed mycobacteria resuspended in mineral oil (Freund's adjuvant), develop a T cell mediated arthritis which is especially prominent in the small joints of the extremities. A T cell line has been established from immunized rats with which the disease can be transferred to irradiated, syngeneic recipients [27]. A subclone of this cell line was shown to respond not only to mycobacteria but also to cartilage proteoglycans, suggesting the presence of a shared T cell epitope [28]. The major T cell epitope in mycobacteria was found to reside in a nonapeptide of a 65K heat-shock protein [29], which showed a partial homology with a link protein of proteoglycans [26]. It is tempting to speculate that the arthritogenic effects of these T cells were mediated through recognition of an homologous epitope in cartilage, e.g. the homologous epitope on the link protein. However, this remains to be demonstrated.

Diseases in Man

Rheumatic fever following infection with streptococci has long been associated with molecular mimicry [30]. Group A streptococci contain antigens that are cross-reactive with tissue in the heart, joints and skin. Cross-reactive antibodies are present in sera from patients with rheumatic fever at the time of onset of diseases, they disappear during quiescent periods and reappear at disease recurrences [31]. In early studies, it was demonstrated that antibodies from rabbits, immunized with bacterial membranes, cross-reacted with human cardiac, skeletal and smooth muscle of blood vessels, but they did not distinguish between rheumatic and non-rheumatic heart tissue [32]. Monoclonal antibodies, obtained from mice immunized with isolated streptococcal membranes, bound to the C terminal part of the streptococcal M protein and cross-reacted with cardiac myosin [33–35]. The pathogenesis of rheumatic fever involves both humoral and cellular autoimmune mechanisms and there are also indications of cross-reactivities at the T cell level, since peripheral blood lymphocytes, activated by streptococcal M protein are cytotoxic to cultured human cells, including heart cells [36]. However, factors other than antigenic mimicry between streptococci and tissue antigens are likely to determine whether the disease develops, since heart reactive antibodies are found after uncomplicated streptococcus infections, although at lower titres. Taking these observations into account, it has recently been suggested that development of the disease is determined not only by cross-reactions between the bacteria and host cell proteins, but also in the particular host by a unique immune response to the self-antigens shared with the microbe and the genetic predisposition to disease [37].

An analogous situation is encountered in Chagas' disease following infection with the parasite *Trypanosoma cruzi.* Chronic Chagas' disease is characterized by destruction of neuronal tissue, accompanied by intense inflammatory reactions and denervation of the heart [38]. Autoimmune reactions were suggested to be involved in the pathogenesis of this disease, since sera from affected individuals contained autoantibodies to normal heart and nerve tissue components, including the endocardium, vascular structures, interstitia of cardiac muscles [39], Schwann's cells [40], laminin [41], striated muscle [42], and neurons [43]. These autoantibodies have been speculated to emerge in response to antigenic mimicking epitopes, which have been identified with monoclonal antibodies that cross-react with trypanosoma and cardiac cells and neurons [44,45]. However, not only direct molecular mimicry between the parasite and uninfected cells could induce cross-reactivity, but parasitic material released from lysed host cells might also adsorb to surrounding host cells and thus initiate a cross-reactive immune response [46]. Definite proof of the autoimmune nature of chronic Chagas' disease is not yet available and it is evident that factors other than antigenic similarities are important, since only a relatively small proportion of patients infected with *Trypanosoma cruzi* progress to overt chronic Chagas' disease.

Coeliac disease is activated by dietary exposure to gliadin, which is a component of wheat gluten, and is characterized by small intestinal mucosal injury and malabsorption [47]. The disease is associated with HLA class II genes, but factors besides genetics influence the pathogenesis, since the disease is not always concordant among monozygotic twins. Therefore, it has been speculated that environmental factors, other than gliadin, may be important in initiating the disease [48]. In an attempt to identify an infectious aetiologic agent, Kagnoff and co-workers searched for sequences shared between a-gliadin and virus proteins [49]. This revealed a region of homology (8 amino acids shared out of 11) between a-gliadin and the Elb protein of human adenovirus Ad12, a virus which can usually be isolated from the human intestinal tract. Antibodies, specific for the Elb protein from Ad12 adenovirus, cross-reacted with the intact a-gliadin molecule, a peptide fragment of the molecule and a synthetic peptide of a-gliadin containing a shared sequence. Evidence for cell mediated immunity to the homologous a-gliadin virus sequence was subsequently obtained from indirect leukocyte migration inhibition assays. Patients with coeliac disease treated with a gluten-free diet showed significantly higher responses to a 12-amino-acid-residue-long peptide containing the shared sequence, than did healthy subjects [50]. Additional observations linking adenovirus Ad12 to the development of coeliac disease include: (a) a higher prevalence of neutralizing antibodies to the Ad12 adenovirus in patients with untreated coeliac disease than in relevant controls, even though this did not reflect a general increase in antibody titres to other viruses in the patients [51], and (b) serum from patients contained antibodies that reacted to a synthetic peptide representing the shared sequence, but not to irrelevant peptides – only patients with active coeliac disease had peptide reactive antibodies, treated patients did not [48,51]. These observations of antigenic mimicry together with a strong association between coeliac disease and HLA class II genes [52], suggest that peptides shared by a-gliadin and a virus encoded protein cross-react at the T cell level, leading to generation of CD4+ T cells which could help in the induction of antibodies to the shared antigenic determinants. Such antibodies could in turn mediate intestinal injury by complement dependent or antibody dependent cellular cytotoxicity [48].

Myasthenia gravis (MG) is caused by an autoimmune mediated reduction in the numbers of nicotinic acetylcholine receptors (AChR), and is clinically characterized by a generalized muscular weakness [53]. The pathogenic mechanisms have been well documented and AChR is one of the best characterized autoantigens. However, little is known about what initiates the autoimmune response to the receptor [2]. Viruses have been speculated to be one of the possible aetiological factors, but there is no direct evidence to support this [54,55]. To test the hypothesis that a virus could induce a cross-reactive autoimmune response to the receptor, we searched for virus protein sequences shared with the α-chain of the AChR. The α-chain is known to represent the most immunogenic part of the receptor, and it has been suggested that the majority of autoantibodies in MG are directed to a major immunogenic region of the α-chain, but this is a matter of debate [2]. We selected a sequence in a highly hydrophilic region of the AChR α-chain (residues 160–167) for analysis [56] and our computer search identified several viral proteins that showed homology with the receptor sequence, including two sequences of the herpes simplex virus (HSV) glycoprotein D [13]. Analysis of sera from rabbits, immunized with synthetic peptides representing the homologous sequences, demonstrated that one of the

HSV glycoprotein D sequences was highly cross-reactive to the AChR sequence. To test if this shared sequence was immunogenically recognized in MG we analysed sera from 40 patients, and found that 6 contained antibodies that bound to a synthetic peptide representing the AChR α-chain sequence [57]. Peptide reactive autoantibodies were affinity purified from the sera of MG patients, yielding IgG antibodies that bound to native AChR and inhibited the binding of α-bungaro-toxin to the receptor. The cross-reactivity of the affinity purified antibodies was demonstrated, as they stained HSV infected cells but not cells infected with human cytomegalovirus or lymphocytic choriomeningitis virus [57]. These data suggest that immunological cross-reactivity to an epitope, shared between a common human virus and AChR, may be associated with initiating some cases of MG.

In another approach to investigate for an infectious aetiology to MG, Stefansson and co-workers used monoclonal antibodies to AChR in western blot analysis of bacterial proteins, to test for shared epitopes [58]. Two out of 50 monoclonal antibodies directed against the AChR α-subunit reacted with proteins from *Escherichia coli*, *Klebsiella pneumoniae*, *Yersinia enterocolitica* and *Proteus vulgaris*. Further, sera from a few patients with MG reacted with two proteins from *Y. enterocolitica*, but it was not known if these antibodies also bound AChR. Taken together, these observations on molecular mimicry between microbial proteins and AChR suggest that MG may be initiated by several infectious aetiological factors in susceptible individuals.

Insulin-dependent, type 1 diabetes mellitus is caused by a selective destruction of β cells, associated with autoimmune phenomena such as mononuclear cell infiltration in the pancreatic islets and autoantibodies to islet cells, specifically reacting with the plasma membrane of β cells [59]. As in the other diseases that have been reviewed here, environmental and genetic factors both appear to be of importance in the pathogenesis. Viruses have been speculated to be one of the aetiological factors, thus there are several examples of acute coxsackie virus infection resulting in type 1 diabetes in man [60,61] and persistent infection with cytomegalo [62] and rubella virus [63] has been associated with development of the disease. Molecular mimicry between a virus antigen and a target cell autoantigen represents only one mechanism whereby viruses may cause autoimmunity. An alternative hypothesis is that autoimmunity could result from lack of tolerance, perhaps mediated by a defective T suppressor cell function. Since the induction and maintenance of T suppressor cells are dependent on presentation of antigen fragments by HLA-DR/DQ molecules [64], the function of these cells would be compromised if there was an immune reaction against the HLA class II molecules themselves. Thus, antigenic mimicry between virus proteins and HLA class II molecules could result in cross-reactive virus antibodies interfering with the development or differentiation of T suppressor cells.

Recently, several examples of shared epitopes between HLA-DR or HLA-DQ β-chain molecules and human virus proteins have been reported. The IE-2 region of human cytomegalovirus was found to share a sequence of five amino acids and show immunological cross-reactivity with a conserved region of the first domain of HLA-DR β-chain molecules [65]. This homology was speculated to present a mechanism, whereby infection with cytomegalovirus might contribute to graft rejection after organ transplantation. In another example, a sequence was identified in the human immunodeficiency virus gp41 envelope protein which showed homology to a conserved N terminal sequence of HLA class II β-chain molecules [66]. Antibody cross-reactivity to the mimicking epitope on HLA class II molecules was suggested to contribute to the functional impairment of CD4+ T cells and development of autoimmune phenomena in AIDS patients, since antibodies from AIDS patients bound to synthetic peptides representing the shared epitope as well as to native HLA class II molecules. These antibodies were further shown to interfere with the cells' function in vitro [67]. The proliferative response of normal CD4+ T cells to tetanus toxoid and allogeneic stimulation was significantly decreased after incubation with patient sera containing HLA class II cross-reactive antibodies, and such sera were shown to destroy HLA class II positive cells by antibody dependent cellular cytotoxicity. In yet another example, sequence homology was demonstrated between the gp110 protein of Epstein–Barr (EB) virus and HLA-DR β-chain antigens positively associated with rheumatic arthritis [68].

HLA-DR4 is in linkage disequilibrium with two alleles of HLA-DQβ, DQw7 and DQw8. These DQ β-chain molecules differ within the first domain by only four amino acids at positions 13, 26, 45 and 57 [69]. The DQw8 allele, and in particular the amino acid in position 57, are closely associated with type 1 diabetes [70]. Recently, we have demonstrated that the particular amino acids in and around position 57 in the β-chain molecules of DQw7 and DQw8 define an epitope which can induce antibodies that specifically distinguish the two closely related proteins [71]. Looking for an infectious pathogen which might initiate type 1 diabetes, we identified a significant amino acid homology, involving 11 out of 17 residues, between the EB virus protein BOLF1 and the diabetes associated HLA-DQw8 β-chain molecule, including the residue in position 57 [72]. Antisera, which were raised to the virus peptide, cross-reacted with the homologous HLA-DQw8 peptide but not with the HLA-DQw7 peptide, only differing by a single amino acid substitution in position 57. Furthermore, antisera to the virus peptide bound specifically to immunoblotted HLA class II β-chain molecules. Since these observations suggested that EB virus might be a predisposing factor for development of type 1 diabetes, sera from children with recent onset of the disease were screened for past or ongoing EB virus infection [73]. There was, however, no correlation between EB virus infection and diabetes, arguing against the idea that EB virus is a principal pathogen in the pathogenesis of type 1 diabetes.

Concluding Remarks

The concept of molecular mimicry has been essential to pharmacology, in the search for substances that differ in chemical structure but display similar biological effects [74]. In immunology, introduction of the molecular mimicry concept was related to parasites' evasion of immune recognition [75,76]. The idea was complementary to that of antigenic mimicry as a cause of autoimmunity, since parasites expressing antigenic determinants mimicking host-determinants were hypothesized not to be recognized by the immune system. Molecular mimicry in this context might not only apply to antigens intrinsic to the parasite, but also to host cell antigens acquired by parasites during infection [77].

The appearance of autoantibodies following infection need not be due to antigenic mimicry between an infecting micro-organism and host cells. Susceptible strains of mice infected with coxsackie B3 virus develop autoimmune myocarditis and autoantibodies specific for cardiac myosin. The myosin autoantibodies, however, do not cross-react with the virus and it has been speculated that the autoimmune response is secondary to viral destruction of myocardie cells followed by release of cardiac myosin [78]. In contrast to these observations in an experimental model, preliminary data show that coxsackie B3 virus shares significant sequence homology with cardiac myosin and that patients with myocarditis, caused by this virus, have antibodies that bind to synthetic peptides representing the homologous regions (P. L. Schwimmbeck, personal communication).

Over the last few years, many reports have emerged on cross-reactivities and sequence homologies between unrelated microbial and host cell proteins. However, the question remains, whether such cross-reactivities are of any significance in the pathogenesis of autoimmune diseases, or rather appear as a consequence of infection, and to what degree the observations of shared amino acid sequences reflect the emergence of cross-reactive antibodies or T cells following infection in vivo.

The challenge for the future will be to provide evidence that these phenomena play a role in initiating autoimmunity. The hypothesis that a shared antigenic determinant on a virus and an autoantigen can precipitate autoimmunity may be too simplistic, but the presence of such common antigens nevertheless represents an intriguing link between virus and autoimmunity.

References

1. Schwartz RH (1989) Acquisition of immunologic self-tolerance. Cell 57:1073–1081
2. Lindstrom JD, Shelton D, Fujii Y (1988) Myasthenia gravis. Adv Immunol 42:233–283
3. Oldstone MBA (1984) Virus-induced immune complex formation and disease: Definition, regulation, importance. In: Notkins AL, Oldstone MBA (eds) Concepts in viral pathogenesis, Springer, Heidelberg, pp 201–209
4. Bjorkman PJ, Saper MA, Samraoui B, Bennett WS, Strominger JL, Wiley DC (1987) The foreign antigen binding site and T cell recognition regions of class I histocompatibility antigens. Nature 329:512–518
5. Davis MM, Bjorkman PJ (1988) T-cell antigen receptor genes and T-cell recognition. Nature 334:395–402
6. Svejgaard A, Platz P, Ryder LP (1983) HLA and disease 1982: a survey. Immunol Rev 70:193–218
7. Michelsen B, Dyrberg T, Vissing H, Serup P, Lernmark Å (1990) HLA-DQ and -DX genes in insulin-dependent diabetes. Curr Top Microbiol Immunol (in press)
8. Olmos P, A'Hern R, Heaton DA, Millward BA, Risley D, Pyke DA, Leslie RDG (1988) The significance of the concordance rate for Type 1 (insulin-dependent) diabetes in identical twins. Diabetologia 31:747–750
9. Notkins AL, Onodera T, Prabhakar B (1984) Virus-induced autoimmunity. In: Notkins AL, Oldstone MBA (eds) Concepts in viral pathogenesis, Springer, Heidelberg, pp 210–215
10. Damian RT (1964) Molecular mimicry: antigen sharing by parasite and host and its consequences. Am Nat 98:129–149
11. Bækkeskov S, Landin M, Kristensen JK, Srikanta S, Bruining GJ, Mandrup-Poulsen T, de Beaufort C, Soeldner JS, Eisenbarth G, Lindgren F, Sundquist G, Lernmark Å (1987)

Antibodies to a 64,000 M_r human islet cell antigen precede the clinical onset of insulin-dependent diabetes. J Clin Invest 79:926–934
12. Dyrberg T, Petersen JS, Oldstone MBA (1990) Immunological cross-reactivity between mimicking epitopes on a virus protein and a human autoantigen depends on a single amino acid residue. Clin Immunol Immunopathol 54:290–297
13. Dyrberg T, Oldstone MBA (1986) Peptides as probes to study molecular mimicry and virus-induced autoimmunity. Curr Topic Microbiol Immunol 130:25–37
14. Lane DP, Hoeffler WK (1980) SV 40 large T shares an antigenic determinant with a cellular protein of molecular weight 68,000. Nature 288:167–170
15. Dales S, Fujinami RS, Oldstone MBA (1983) Infection with vaccinia favors the selection of hybridomas synthesizing autoantibodies against intermediate filaments, one of them cross-reacting with the virus hemagglutinin. J Immunol 131:1546–1553
16. Fujinami RS, Oldstone MBA, Wroblewska Z, Frankel ME, Koprowski H (1983) Molecular mimicry in virus infection: cross-reaction of measles virus phosphoprotein or of herpes simplex protein with human intermediate filaments. Proc Natl Acad Sci USA 80:2346–2350
17. Tardieu M, Powers ML, Hafler DA, Hauser SL, Weiner HL (1984) Autoimmunity following viral infection: demonstration of monoclonal antibodies against normal tissue following infection of mice with reovirus and demonstration of shared antigenicity between virus and lymphocytes. Eur J Immunol 14:561–565
18. Srinivasappa J, Saegusa J, Prabhakar BS, Gentry MK, Buchmeier MJ, Wiktor TJ, Koprowski H, Oldstone MBA, Notkins AL (1986) Molecular mimicry: frequency of reactivity of monoclonal antiviral antibodies with normal tissues. J Virol 57:397–401
19. Oldstone MBA (1989) Molecular mimicry as a mechanism for the cause and as a probe uncovering etiologic agent(s) of autoimmune disease. Curr Top Microbiol Immunol 145:127–135
20. Fujinami RS, Oldstone MBA (1985) Amino acid homology and immune responses between the encephalitogenic site of myelin basic protein and virus: a mechanism for autoimmunity. Science 230:1043–1045
21. Jahnke U, Fischer EH, Alvord EC (1985) Sequence homology between certain viral proteins and proteins related to encephalomyelitis and neuritis. Science 229:282–284
22. Shaw S-Y, Laursen RA, Lees MB (1986) Analogous amino acid sequences in myelin proteolipid and viral proteins. FEBS 207:266–270
23. Lipton H, Miller S, Melvold R, Fujinami RS (1986) Theiler's murine encephalomyelitis virus (TMEV) infection in mice as a model for multiple sclerosis. In: Notkins AL, Oldstone MBA (eds) Concepts in viral pathogenesis II. Springer-Verlag, Heidelberg, pp 248–254
24. Fujinami RS, Zurbriggen A, Powell HC (1988) Monoclonal antibody defines determinant between Theiler's virus and galactocerebroside. J Neuroimmunol 20:25–32
25. Fujinami RS (1989) Immune responses against myelin basic protein and/or galactocerebroside cross-react with viruses: Implications for demyelinating disease. Curr Top Microbiol Immunol 145:91–100
26. van Eden W, Hogervorst EJM, Hensen EJ, van der Zee R, van Embden JDA, Cohen IR (1989) A cartilage-mimicking T-cell epitope on a 65k mycobacterial heat-shock protein: adjuvant arthritis as a model for human rheumatoid arthritis. Curr Top Microbiol Immunol 145:27–43
27. Holoshitz J, Naparstek Y, Ben-Nun A, Cohen IR (1983) Lines of T lymphocytes induce or vaccinate against autoimmune arthritis. Science 219:56–58
28. van Eden W, Holoshitz J, Nevo Z, Frenkel A, Klajman A, Cohen IR (1985) Arthritis induced by a lymphocyte clone that responds to *Mycobacterium tuberculosis* and to cartilage proteoglycans. Proc Nat Acad Sci USA 82:5064–5067
29. van der Zee R, van Eden W, Meloen RH, Noordzÿ A, van Embden JDA (1989) Efficient mapping and characterization of a T cell epitope by the simultaneous synthesis of multiple peptides. Eur J Immunol 19:43–47
30. Zabriskie JB (1985) Rheumatic fever: the interplay between host genetics and microbe. Circulation 71:1077–1086
31. Zabriskie JB, Hsu KC, Seegal BC (1970) Heart reactive antibody associated with rheumatic fever: characterization and diagnostic significance. Clin Exp Immunol 7:149–159
32. Zabriskie JB, Freiner EH (1966) An immunological relationship between the group A streptococcus and mammalian muscle. J Exp Med 124:661–678
33. Dale JB, Beachey EH (1985) Epitopes of streptococcal M proteins shared with cardiac myosin. J Exp Med 162:583–591
34. Krisher K, Cunningham MW (1985) Myosin: a link between streptococci and heart. Science 227:413–415
35. Sargent SJ, Beachey E, Corbett CE, Dale JB (1987) Sequence of protective epitopes of

streptococcal M proteins shared with cardiac sarcolemmal membranes. J Immunol 139, 1285–1290
36. Dale JB, Beachey EH (1987) Human cytotoxic T-lymphocytes evoked by group A streptococcal M proteins. J Exp Med 166:1825–1835
37. Froude J, Gibofsky A, Buskirk DR, Khanna A, Zabriskie JB (1989) Cross-reactivity between streptococcus and human tissue: A model of molecular mimicry and autoimmunity. Curr Top Microbiol Immunol 145:5–26
38. Takle GB, Hudson L (1989) Autoimmunity and Chagas' disease. Curr Top Microbiol Immunol 145:79–92
39. Cossio PM, Diez C, Szarfman A, Kreutzner E, Candiolo B, Arana RM (1974) Chagasic cardiomyopathy: demonstration of a serum gamma globulin factor which reacts with endocardium and vascular structures. Circulation 49:13–25
40. Khoury EL, Ritacco V, Cossio PM, Laguens RP, Szarfman A, Diez C, Arana RM (1979) Circulating antibodies to peripheral nerve in American trypanosomiasis (Chagas' disease). Clin Exp Immunol 36:8–15
41. Szarfman A, Luguettia A, Rossi A, Rezende JM, Schmunis GA (1981) Tissue reacting immunoglobulins in patients with different clinical forms of Chagas' disease. Am J Trop Med Hyg 30:43
42. Cossio PM, Laguens RP, Kreutzer E, Diez C, Segal A, Arana RM (1974) Chagasic cardiopathy: antibodies reacting with plasma membrane of striated muscle and endothelial cells. Circulation 50:1252–1263
43. Ribiero dos Santos R, Marquez JO, Von Gal Furtado CC, Ramos de Olieveira JC, Martins AR, Koberle F (1979) Antibodies against neurons in chronic Chagas' disease. Tropenmed Parasitol 30:19–23
44. Wood JN, Hudson L, Jessell TM, Yamamoto M (1982) A monoclonal antibody defining determinants on subpopulations of mammalian neurons and *Trypanosoma cruzi* parasites. Nature 296:34–38
45. Towbin H, Rosenfielder G, Wieslander J, Avila JL, Rojas M, Szarfman A, Esser K, Nowack H, Timpl R (1987) Circulating antibodies to mouse laminin in Chagas' disease. American cutaneous leishmaniasis and normal individuals recognize terminal galactosyl α(1–3)-galactose epitopes. J Exp Med 166:419–432
46. Ribiero dos Santos R, Hudson L (1980) Immunological consequences of parasite modification of host cells. Clin Exp Immunol 40:36–41
47. Kagnoff MF (1990) Celiac disease: pathogenesis and clinical features. In: Thomson ABR, Shaffer EA (eds) Modern concepts in gastroenterology. Plenum, New York (in press)
48. Kagnoff MF (1989) Celiac disease: Adenovirus and alpha gliadin. Curr Top Microbiol Immunol 145:67–78
49. Kagnoff MF, Austin RK, Hubert JJ, Bernadin JE, Kasarda DD (1984) Possible role for a human adenovirus in the pathogenesis of celiac disease. J Exp Med 160:1544–1557
50. Karagiannis JA, Priddle JD, Jewell DP (1987) Cell-mediated immunity to a synthetic gliadin peptide resembling a sequence from adenovirus 12. Lancet ii:884–886
51. Kagnoff MF (1987) Evidence for the role of a human intestinal adenovirus in the pathogenesis of celiac disease. Gut 28:995–1001
52. Howell MD, Smith JR, Austin RK, Kelleher D, Nepom GT, Volk B, Kagnoff MF (1988) An extended HLA-D region haplotype associated with celiac disease. Proc Natl Acad Sci USA 85:222–226
53. Drachman D (1978) Myasthenia gravis. N Engl J Med 298:136–186
54. Aoki T, Drachman DB, Asher DM, Gibs CJ, Bahmanyar S, Molinsky JS (1985) Attempts to implicate viruses in myasthenia gravis. Neurology 35:185–192
55. Klavinskis LS, Willcox HN, Richmond JE, Newsom-Davis J (1986) Attempted isolation of viruses from myasthenia gravis thymus. J Neuroimmunol 11:287–299
56. Noda M, Furutani Y, Takahashi H, Toyosato M, Tanabe T, Shimizu S, Kikyotani S, Kayano T, Hirose T, Inayama S, Numa S (1983) Cloning and sequence analysis of calf cDNA and human genomic DNA encoding alpha-subunit precursor of muscle acetylcholine receptor. Nature 305:818–823
57. Schwimmbeck PL, Dyrberg T, Drachman DB, Oldstone MBA (1989) Molecular mimicry and myasthenia gravis: an autoantigenic site of the acetylcholine receptor α subunit that has biologic activity and reacts immunochemicaly with herpes simplex virus. J Clin Invest 84:1174–1181
58. Stefansson K, Dieperink ME, Richman DP, Gomez CM, Marton LS (1985) Sharing of antigenic determinants between the nicotenic acetylcholine receptor and proteins in *Escherichia coli*, *Proteus vulgaris* and *Klebsiella pneumoniae*. N Engl J Med 312:221–225

59. Lernmark Å (1985) Molecular biology of type 1 (insulin-dependent) diabetes mellitus. Diabetologia 28:195–203
60. Yoon JW, Austin M, Onodera T, Notkins AL (1979) Virus induced diabetes mellitus: Isolation of a virus from the pancreas of a child with diabetic ketoacidosis. N Engl J Med 300:1173–1179
61. Champsaur H, Dussaix E, Samolyk D, Fabre M, Bach CH, Assan R (1980) Diabetes and coxsackie virus B 5 infection. Lancet i:251
62. Pak CY, Eun HM, McArthur RG, Yoon JW (1988) Association of cytomegalovirus infection with autoimmune Type 1 diabetes. Lancet ii:1–4
63. Ginsberg-Fellner F, Witt ME, Fedun B, Taub F, Dobersen MJ, McEvoy RC, Cooper LZ, Notkins AL, Rubinstein P (1985) Diabetes mellitus and autoimmunity in patients with the congenital rubella syndrome. Rev Infect Dis 7 (suppl 1):170–176
64. Lamb JR, Feldmann M (1984) Essential requirement for major histocompatibility complex recognition in T-cell tolerance induction. Nature 308:72–74
65. Fujinami RS, Nelson JA, Walker L, Oldstone MBA (1988) Sequence homology and immunologic cross-reactivity of human cytomegalovirus with HLA-DR beta chain: a means for graft rejection and immunosuppression. J Virol 62:1001–1005
66. Golding H, Robey FA, Gates FT, Linder W, Reining PR, Hoffman T, Golding B (1988) Identification of homologous regions in human immunodeficiency virus I gp41 and human MHC class II β I domain. J Exp Med 167:914–923
67. Golding H, Shearer GM, Hillman K, Lucas P, Manischewitz J, Zajac RA, Clerici M, Gress RE, Boswell NR, Golding B (1989) Common epitope in human immunodeficiency virus (HIV) I-GP41 and HLA class II elicits immunosuppressive autoantibodies capable of contributing to immune dysfunction in HIV I-infected indivisuals. J Clin Invest 83:1430–1435
68. Roudier, J, Petersen J, Rhodes GH, Luka J, Carson DA (1989) Susceptibility to rheumatoid arthritis maps to a T-cell epitope shared by the HLA-Dw4 DR β-1 chain and the Epstein–Barr virus glycoprotein gp110. Proc Natl Acad Sci USA 86:5104–5108
69. Michelsen B, Lernmark Å (1987) Molecular cloning of a polymorphic DNA endonuclease fragment associates insulin-dependent diabetes mellitus with HLA-DQ. J Clin Invest 79:1144–1152
70. Todd JA, Bell JI, McDevitt HO (1987) HLA-DO_{β} gene contributes to susceptibility and resistance to insulin-dependent diabetes mellitus. Nature 329:599–604
71. Atar D, Dyrberg T, Michelsen B, Karlsen A, Kofod H, Mølvig J, Lernmark Å (1989) Site-specific antibodies distinguish single amino acid substitutions in position 57 in HLA-DQ β-chain alleles associated with insulin-dependent diabetes. J Immunol 143:533–538
72. Dyrberg T, Michelsen B, Oldstone M (1988) Virus and host cell antigen sharing in myasthenia gravis and autoimmune diabetes. In: Lernmark Å, Dyrberg T, Terenius L, Hökfelt B (eds) Molecular mimicry in health and disease. Excerpta Medica, Amsterdam, pp 245–254
73. Sairenji T, Daibata M, Sorli CH, Qvistbäck H, Humphreys RE, Ludvigsson J, Palmer J, Landin-Olsson M, Sundkvist G, Michelsen B, Lernmark Å, Dyrberg T (1990) Relating homology between the EBV BOLF-1 molecule and HLA-DQw8 β chain to recent onset Type 1 diabetes mellitus (submitted)
74. Burgen ASV (1988) Molecular mimicry and drugs. In: Lernmark Å, Dyrberg T, Terenius L, Hökfelt B (eds) Molecular Mimicry in Health and Disease. Excerpta Medica, Amsterdam, pp 3–11
75. Damian RT (1974) Molecular mimicry: antigen sharing by parasite and host and its consequences. Am Nat 98:129–149
76. Damian RT (1988) Parasites and molecular mimicry. In: Lernmark Å, Dyrberg T, Terenius L, Hökfelt B (eds) Molecular mimicry in health and disease. Excerpta Medica, Amsterdam, pp 211–218
77. Bloom BR (1979) Games parasites play. Nature 279:21–26
78. Neu N, Rose NR, Beisel KW, Herskowitz A, Gurri-Glass G, Craig SW (1987) Cardiac myosin induces myocarditis in genetically predisposed mice. J Immunol 139:3630–3636

Chapter 13

A Heat Shock Protein, Molecular Mimicry and Autoimmunity

I. R. Cohen

Molecular Mimicry and Autoimmunity

The function of the immune system is to gather information about macromolecules entering the individual's biological environment and to deploy various attack mechanisms when the macromolecules signify a threat to the integrity of the body. This function occurs naturally when the immune system responds to an invading bacterium or virus; it occurs iatrogenically when the individual is vaccinated with a microbial product or transplanted with a foreign tissue graft; and it occurs pathologically in autoimmune disease when the immune system turns against a normal constituent of the body as if it were a foreign menace. To avoid autoimmune disease, therefore, the immune system must be able to sort out the foreign from the self; it must be able to make decisions based on the molecular information it has gathered.

The task of separating the foreign from the self is complicated by the fact that the basic unit of immunological information, the epitope, is only a relatively small part of a macromolecule on any invading micro-organism. The 6–10 amino acids presented to T cells by major histocompatibility complex (MHC) molecules or bound by the combining site of an antibody molecule constitute important, if not decisive immunological information. Such a tiny bit of a single macromolecule can serve as the target of an immune attack destroying a foreign invader or, unfortunately, of an autoimmune attack incapacitating a vital tissue. Thus, health or disease can ride on a triflingly small sequence or structure among the myriads of potential epitopes comprising a cell or virus. To put it another way, the immunological view of the world is modular; the immune system makes decisions on the basis of what fits its receptors.

A danger inherent in the modular view of the world is mimicry between foreign and self; a microbe and the self need share only a small bit of sequence or structure to look alike to the immune system [1]. The outcome of such mimicry could be the triggering of an autoimmune reaction by a microbe.

An important question is whether and how often mimicry between microbe and self actually produces autoimmune disease. Considering the great numbers of potential epitopes in any macromolecule and the great numbers of macromolecules of which parasites and hosts are composed, shared epitopes must be relatively common. Moreover, many antibodies have been found to bind both to microbial and self epitopes, indicating that microbial antibodies may behave as autoantibodies, and vice versa [2]. However, autoimmune reactivity by itself does not necessarily lead to autoimmune disease. Individuals may have antibodies or T cells that can interact with epitopes present in their own tissues without any signs of disease [2]. Indeed, healthy persons have T cells reactive to major pathogenic self-antigens such as myelin basic protein or antibodies reactive to DNA or acetylcholine receptor. These same elements are characteristic of major autoimmune diseases. What, then, determines the pathogenic potential of an autoimmune reaction? When and how does benign autoimmunity evolve into virulent autoimmune disease? At present there is no firm answer to the question. However, the fact that there are few experimental autoimmune diseases inducible by immunization to microbial antigens suggests that the transition from microbial mimicry of self to overt disease is not common.

Adjuvant Arthritis

One of the few examples of autoimmune disease triggered by microbial immunization is adjuvant arthritis (AA). This disease, discovered about 30 years ago by Pearson [3], is a progressive polyarthritis of the small joints of the extremities of rats triggered by immunization to dead *Mycobacterium tuberculosis* (MT) organisms. AA features extra-articular manifestations and has been considered to model aspects of rheumatoid arthritis in humans [4]. It was discovered that AA could be transferred to naive rats using lymphoid cells of affected rats [4], indicating that AA was probably an immunological disease, possibly an autoimmune disease triggered by mimicry between MT and the joints of the rat. Thus AA appeared as a suitable subject to investigate microbial mimicry and autoimmune disease.

A T Cell Strategy

As we have done in other autoimmune diseases [5], my associates and I studied AA by isolating in pure culture the aetiological agent: the rat T cells responsible for recognizing MT and producing arthritis. An arthritogenic T line was isolated using whole, pulverized MT as a selecting antigen [6]. This line was cloned and an arthritogenic clone, A2b, was isolated [7]. This rat T cell responsive to MT and capable of causing arthritis in rats was therefore an ideal probe to decipher the modular arthritogenic epitope shared by MT and joints.

Two results were obtained by assaying the reactivities of A2b to either MT antigens or to materials isolated from joints. On the side of the joints, A2b was observed to recognize crude protein extracts of cartilage proteoglycan [8]. On the side of MT, A2b was found to recognize an acetone-perceptible fraction (AP) of MT [9]. Following the results in rats, T cell reactivity to the AP fraction of MT was also detected among the synovial fluid T cells of humans with active rheumatoid arthritis [9]. Thus, mimicry between an epitope in MT and cartilage proteoglycan was associated with the aetiology of AA and might also be a factor in rheumatoid arthritis. What was this epitope?

A partial answer and a number of new and more perplexing questions next resulted from the use of the tools of genetic engineering to pinpoint the A2b epitope on MT.

The 65 kDa Heat Shock Protein

To aid investigation of the bacteriology and immunology of mycobacterial infections, researchers have engineered the expression of mycobacterial genes in *E. coli* [10]. It was therefore possible to use ready-made gene libraries to screen for the reactivity of T clone A2b. The outcome was the identification of the critical MT antigen as the 65 kDa heat shock protein (hsp65) and its specific epitope as the nine amino acids at positions 180–188 [11]. The nonapeptide sequence of hsp65, TFGLQLELT, was likely to be involved in AA because A2b, the T cell responsive to it, was involved in AA, both as a cause of arthritis [7] and as a therapeutic agent functional in T cell vaccination against AA [12]. The association of hsp65 with arthritis was further strengthened by the observation that T cells of rheumatoid arthritis patients also respond to hps65 [13].

Where is the Arthritogenic Mimicry?

How then might a T cell immune response to hsp65 and its 180–188 peptide cause arthritis? Computer searches of protein sequence banks for proteins with homology with the 180–188 peptide sequence have not been rewarding, except for a weak, four amino acid homology with a sequence of the link protein of the rat proteoglycan, TAVVALELE [14]. However, clone A2b does not respond to this link protein peptide in vitro (W. van Eden et al., personal communication; D. Markovits et al., personal communication). Thus, either the link protein sequence is recognized by A2b in vivo but not in vitro, or the mycobacterial 180–188 sequence appears on another, unknown joint molecule. The fact that A2b responded to crude preparations of cartilage suggests that a target molecule with the 180–188 sequence should be identifiable.

The 180–188 sequence was found to be a target of both T cell reactivity and antibodies in the peripheral blood of children with juvenile rheumatoid arthritis (M. G. Danieli et al., personal communication). The affected children also responded to the link peptide sequence. However, adults with rheumatoid arthritis did not seem to respond to either peptide, although they did respond to the whole hsp65 molecule. Thus, the 180–188 peptide sequence of hsp65 may be important in some forms of human arthritis as well as in AA. What about other epitopes of hsp65?

Hsp molecules are among the most conserved of proteins and mycobacterial and mammalian hsp65 show about 50% amino acid homology [15]. Thus the mammalian hsp65 molecule itself might conceivably serve as the target for an arthritogenic response induced by mycobacterial hsp65. However, the expression of hsp molecules can be induced by a variety of stresses in most types of cells and there is no obvious reason why the mammalian hsp65 molecule should be specific for the joints. Perhaps the stresses of normal use induce a constitutive expression of hsp65 in the joints so that anti-hsp65 T cells might find a target in the joints more easily than in other tissues. Once arthritis was initiated the continued expression of augmented amounts of hsp65 due to inflammation could function to maintain the arthritogenic process.

However, even if mammalian hsp65 were to be expressed in the joints of rats or of children with juvenile rheumatoid arthritis, the mammalian hsp65 molecule of known sequence differs markedly from mycobacterial hsp65 at positions 180–188 [15]. Hence, the anti-180–188 responses associated with these forms of arthritis would not react to the known mammalian sequence. In fact, A2b does not react to human hsp65 (D. Markovits, personal communication). Hence the mammalian molecule that mimics the 180–188 sequence of mycobacterial hsp65 is not likely to be the mammalian hsp65 molecule. Perhaps a joint-specific tissue isotype of hsp65 does carry the 180–188 sequence.

Another problem with the connection between hsp65 and arthritis is that AA has not been inducible in our hands using purified mycobacterial hsp65. Injection of rats with mycobacterial hsp65 in oil is not arthritogenic; it only makes rats refractory to AA [11]. It may be claimed that resistance to AA is also a meaningful connection between hsp65 and disease, but the cause of disease is not simply an immune response to hsp65.

We have isolated another clone of rat T cells from rats with AA, designated M1 (D. Markovits, personal communication). Clone M1 recognizes a peptide sequence of hsp65 outside of the 180–188 sequence recognized by A2b. As I shall describe below, M1 seems to bear a major idiotype involved in the regulation of AA and the properties of its epitope are now under investigation.

Hsp65 is Immunologically Dominant

An unexpected feature of hsp65 is its immunological dominance. Individuals with tuberculosis or leprosy respond preferentially to hsp65 [16,17]. This is perplexing. According to traditional thinking, molecules rich in self-epitopes ought to be weak immunogens because of self-tolerance. Mycobacterial hsp65, as we pointed out above, has 50% amino acid homology with human hsp65, including sizeable stretches of identify. The association of hsp65 with AA and with rheumatoid arthritis attests to the danger of its autoimmunogenic potential. Why then do the immune systems of persons with mycobacterial infections select hsp65 as a major target? Shouldn't self-tolerance to the shared epitopes prevent an immune response? Surely, mycobacteria bear other molecules with much less similarity to self, and hence with much less autoimmunogenic potential.

Regulation of Anti-hsp65 T Cells

It is unlikely that the immunological dominance of hsp65 can be explained by preferential binding of hsp65 peptides to class II MHC molecules; hsp65 is immunologically dominant in patients irrespective of their MHC genotype (M. G. Danieli et al., personal communication), as it is dominant in rats.

Another possibility is that the dominance of hsp65 is encoded in an anti-idiotypic network of regulatory T cells [18,19]. We find that healthy rats possess helper T cells specifically reactive to the M1 T cell idiotype (N. Karin and I. R. Cohen, unpublished work). These anti-M1 anti-idiotypic helper T cells become markedly activated within several days of immunizing rats with whole MT. Thus it is conceivable that the anti-hsp65 T cells are driven by these preformed anti-idiotypic T cells along with the hsp65 antigen present in MT. Hence the hsp65 reactive T cells may enjoy an advantage over T cells responsive to other MT antigens for which there are no anti-idiotypic helpers. The immunological dominance of the hsp65 antigen is associated with a pre-formed regulatory network [18].

Obviously, the discovery of a natural anti-M1 anti-idiotypic network only raises more questions. What is the origin of the anti-idiotypic network? What purpose does it serve?

It now appears that many humans possess T cells reactive to the amino acid sequences of mycobacterial hsp65 that are homologous to human hsp65 sequences (S. H. Kaufmann, personal communication). The presence of the anti-hsp65 autoreactive T cells is not a sign of autoimmune disease; these humans are healthy. Whatever their purpose, the anti-hsp65 T cells in humans could activate the anti-idiotypic T cells that we believe generate the dominance of the hsp65 antigen.

In addition to helper T cells specific for the M1 anti-hsp65 T cells, we find that rats also possess anti-idiotypic suppressor T cells specific for the M1 anti-hsp65 T cells (N. Karin and I. R. Cohen, unpublished work). Indeed we have found that clones of anti-idiotypic helper and suppressor T cells also regulate autoimmune encephalomyelitis [20].

Combining these observations, one may consider the idea that the anti-idiotypic suppressor cells function to inhibit arthritis produced by the anti-hsp65 T cells [21]. Thus, the network that makes the hsp65 a dominant antigen (the anti-idiotypic helpers) also functions to ensure that the response is well controlled (the anti-idiotypic suppressors). It is quite possible that the predictable dominance of hsp65 spares the individual from making unforeseen and, therefore poorly controlled autoimmune responses [19]. Thus the immunological dominance of hsp65 and the other major autoantigens such as myelin basic protein or insulin may function in protecting the majority against autoimmune disease. The price for this protection is paid by the minority of persons who fail to regulate the dominant responses and thus suffer from autoimmune diseases to these same dominant autoantigens.

Mimicry in Reverse

This chapter began by considering the thesis that microbes may activate an autoimmune response to self due to their molecular mimicry of self-epitopes. Investigation of the regulation of the immune response to hsp65, however, leads us to consideration of the complementary thesis: regulatory networks organized to focus and thus control autoimmune reactions may in turn influence the dominant focus of the immune response to microbial antigens. Could it be that mycobacterial hsp65 is the dominant "foreign antigen" in tuberculosis and leprosy precisely because this molecule draws the attention of a pre-formed T cell network to its self-epitopes? The network-encoded dominance of the self-epitopes "helps" to establish the dominance of the adjacent foreign epitopes. Microbial hsp65 is dominant, not in spite of self mimicry but because of self mimicry.

References

1. Damian RT (1988) Parasites and molecular mimicry. In: Lernmark Å, Dyrberg T, Terenius L, Hökfelt B (eds) Molecular mimicry in health and disease. Elsevier Science, Amsterdam, pp 211–218
2. Shoenfeld Y, Cohen IR (1987) Infection and autoimmunity. In: Sela M (ed) The antigens, vol VII. Academic Press, San Diego New York, pp 307–325
3. Pearson CM (1956) Development of arthritis, periarthritis and periostitis in rats given adjuvant. Proc Soc Exp Biol Med 91:95–101
4. Pearson CM (1964) Experimental models in rheumatoid disease. Arthritis Rheum 7:80–86
5. Cohen IR (1986) Regulation of autoimmune disease: physiological and therapeutic. Immunol Rev 94:5–21
6. Holoshitz J, Naparstek Y, Ben-Nun A, Cohen IR (1983) Lines of T lymphocytes induce or vaccinate against autoimmune arthritis. Science 219:56–58
7. Holoshitz J, Matitiau A, Cohen IR (1984) Arthritis induced in rats by clones of T lymphocytes responsive to mycobacteria but not to collagen type II. J Clin Invest 73:211–215
8. van Eden W, Holoshitz J, Nevo Z, Frenkel A, Klajman A, Cohen IR (1985) Arthritis induced by a T lymphocyte clone that responds to *Mycobacterium tuberculosis* and to cartilage proteoglycans. Proc Natl Acad Sci USA 82:5064–5067
9. Holoshitz J, Klajman A, Drucker et al. (1986) T-lymphocytes of rheumatoid arthritis patients show augmented reactivity to a fraction of mycobacteria cross-reactive with cartilage. Lancet ii:305–309
10. Thole JER, Dauwerse HG, Das RK, Groothuis DG, Schouls LM, Van Embden JDA (1985) Cloning of *Mycobacterium bovis* BCG DNA and expression of antigens in *E. coli*. Infect Immun 50:800–806
11. van Eden W, Thole JER, van der Zee R et al. (1988) Cloning of the mycobacterial epitope recognized by T lymphocytes in adjuvant arthritis. Nature 31:171–173
12. Lider O, Karin N, Shinitzky M, Cohen IR (1987) Therapeutic vaccination against adjuvant

arthritis using autoimmune T lymphocytes treated with hydrostatic pressure. Proc Natl Acad Sci USA 84:4577–4580
13. Res PCM, Schaar CG, Breedveld FC et al. (1988) Synovial fluid T cell reactivity against 65 kD heat shock protein of mycobacteria in early chronic arthritis. Lancet ii:478–480
14. Cohen IR (1988) The self, the world and autoimmunity. Sci Am 258:52–60
15. Jindal S, Dudani AK, Harley CB, Singh B, Gupta RS (1989) Primary structure of a human mitochrondrial protein homologous to the bacterial and plant chaperonins and to the 65-kilodalton mycobacterial antigen. Molec Cell Biol 9:2279–2283
16. Young DB, Ivanyi J, Cox JH, Lamb JR (1987) The 65kDa antigen of mycobacteria: a common bacterial antigen. Immunol Today 8:215–219
17. Young D, Lathigra R, Hendrix R, Sweetser D, Young RA (1988) Stress proteins as immune targets in leprosy and tuberculosis. Proc Natl Acad Sci USA 85:4267–4270
18. Cohen IR (1990) Natural id-anti-id networks and the immunological homunculus. In: Atlan H, Cohen IR (eds) Theories of immune networks. Springer, Berlin, pp 6–12
19. Cohen IR (1990) T-cell vaccination and suppression of autoimmune disease. In: Melchers F et al. (eds) Progress in Immunology VII. Springer, Berlin, pp 867–873
20. Lider O, Reshef T, Beraud E, Ben-Nun A, Cohen IR (1988) Anti-idiotypic network induced by T cell vaccination against experimental autoimmune encephalomyelitis. Science 239:181–183
21. Cohen IR, Atlan H (1990) Network regulation of autoimmunity: an automaton model. J Autoimmunity 2:211–218

Chapter 14

What Do Human T Cells See in the Mycobacterial 65 kDa Heat Shock Protein?

R. R. P. de Vries, W. C. A. van Schooten and
T. H. M. Ottenhoff

Introduction and Summary

Heat shock proteins are ubiquitous proteins that have a crucial function in the biology of both pro- and eukaryotic cells [1]. Their general function seems to be in protein assembly and transport [2,3]. Recent evidence indicates that at least one of the heat shock proteins (hsp) is also involved in antigen binding [4] and the gene coding for this particular heat shock protein (a 70 kDa protein) in humans is located in the HLA system [5].

These proteins are extremely well conserved in evolution and several microbial hsp thus show extensive molecular mimicry with the homologous hsp of their host [2,6]. Actual cross-reactivity between T cells reactive with bacterial hsp and the homologous human hsp has been documented [7,8]. Such cross-reactivity may be detrimental as is dramatically illustrated by the adjuvant arthritis model in rats induced by T cells reactive with the 65 kDa hsp of *Mycobacterium tuberculosis* (see Chapter 13). Nevertheless heat shock proteins appear to be very immunodominant antigens [9]. Immunologically the best studied hsp are mycobacterial hsp's. In both mouse and man they are immunodominant for both B and T cells, both for T cells with an $\alpha\beta$ and those with a $\gamma\delta$ T cell receptor [10,11]. Irun Cohen has addressed this interesting paradox in the previous chapter and concluded that this may be because of and not in spite of self-mimicry.

In this last chapter we will return to the major histocompatibility complex (MHC) (see Chapter 1) and its role in the regulation of the immune response, in particular in preventing immunopathology. Using the 65 kDa hsp as an example we will ask and try to answer a series of questions relating to this issue. According to Irun Cohen the focus of the immune system towards hsp is towards the whole molecule (see Chapter 13). Assuming that he is right, we were interested to know whether the human T cell epitopes mapped on the mycobacterial 65 kDa hsp display a bias towards or away from the human 65 kDa hsp. Because self-reactive T cells might have been deleted or inactivated during ontogeny in the thymus [12,13] we predicted that there would be a bias away from self. The answer, however, is that there was no such bias. This suggests that T cells reactive with both mycobacterial and autologous might be present, which appeared to be the case. Next we analysed whether this cross-reactivity might be predicted from the sequence (molecular mimicry) and found that this was indeed the case. The fact that the documented human T cells reactive with self 65 kDa hsp were apparently not causing ill effects implies that such T cells must be regulated by the immune system. That T cells may trigger or cause immunopathology may well be illustrated with the example of leprosy and we will present a hypothesis on the regulatory role of HLA class II immune response (Ir) genes in relation to expression of immunopathology. Furthermore, we studied the role of HLA class II molecules in the regulation of the response to the mycobacterial 65 kDa hsp and show that HLA-DR Ir genes are operative both in the response to the whole protein and to defined epitopes of this molecule. Finally, our data suggest that in the case of the 65 kDa hsp there may be a molecular basis for immunodominance of a particular T cell epitope.

No Bias Towards or Away From Self-Like Epitopes

Figure 14.1 shows a T cell epitope map of the 65 kDa heat shock protein of *Mycobacterium tuberculosis*, which is identical to *Mycobacterium bovis* BCG, and *Mycobacterium leprae*, which in its turn is 95% identical to the *Mycobacterium tuberculosis* 65 kDa heat shock protein [14]. We will confine ourselves to human T cells mainly because they are most relevant to the diseases, such as leprosy and rheumatoid arthritis, that we are interested in. T cell clones and lines have been used as T cell reagents, the majority from tuberculosis of leprosy patients but some also from healthy contacts. The epitopes were defined using these T cell reagents and deletion mutants of the 65 kDa gene [14] or peptides [7,8,15,16,17]. The majority of these data were generated by van Schooten, but the figure also contains data from other investigators, including Johnathan Lamb and Fredrik Oftung. It is clear that the response to almost all the epitopes is restricted by HLA-DR molecules and, interestingly, every epitope is exclusively restricted by one HLA-DR allele. We will come back to that later, but the fact that there is such clear genetic restriction of T cell reactivity to the 65 kDa protein already implies that the 65 kDa hsp is not a so-called superantigen, but at least in this respect behaves like a conventional T cell antigen.

Recently the sequence of the human analogue became available, which is 50% identical and has 20% conservative amino acid substitutions compared to the mycobacterial 65 kDa hsp [6,18]. Therefore we could ask the question: do T cells selected for reactivity with the mycobacterial 65 kDa protein show a bias for self or for non-self parts of the 65 kDa molecule? The data on which we performed that analysis are shown in Table 14.1. The dots indicate amino acid identity. Recall that the whole 65 kDa mycobacterial hsp shows 50% identity with its

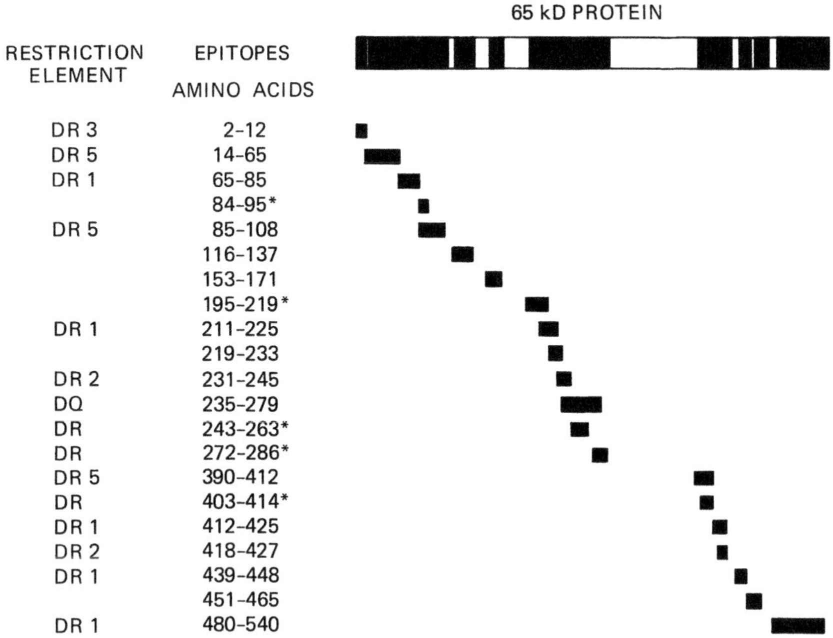

Fig. 14.1. T cell epitope map of the 65 kDa antigen. The T cell epitopes were mapped by reactivity of human T cells to recombinant fusion proteins and synthetic peptides. An *asterisk* marks regions containing self-epitopes identical to the human 65 kDa protein. For a number of T cells the HLA restriction type was established. Fourteen regions were recognized by DR restricted T cells, one region by DQ restricted T cells. For three regions that were recognized by DR restricted T cells, the exact haplotype was not established. As far as studied, each T cell epitope appears to be uniquely associated with a particular restriction element (data from [7,8,14–16,19]).

human homologue. You can count for yourself and then will agree that there is not a bias of focus towards or away from self. In other words, during the development of the T cell repertoire there has not been a trend towards deletion of T cells reactive with the autologous 65 kDa hsp. This implies that we may expect some T cells that are reactive with the mycobacterial 65 kDa hsp to be cross-reactive with the self 65 kDa hsp and other T cells not. That is exactly what has been observed [7,8].

Cross-reactivity With Self May Be Predicted from the Sequence

Five T cell clones reactive with mycobacterial 65 kDa hsp epitopes have been tested for cross-reactivity with the homologous human hsp peptides [7]. Three T cell clones were shown to cross-react with the autologous hsp homologue and two T cell clones did not react with the self homologue (Table 14.1). These T cell clones were from tuberculosis patients and there are no indications that these

Table 14.1. Peptide T cell epitopes of the mycobacterial 65 kDa hsp and mimicry with the human homologue (see [6,7,14–16,18,19])

Species	Epitope	Sequence	T cell clones
M. leprae	3- 13	KTIAYDEEARR	+
M. tuberculosis	3- 13	....	+
Human	28- 38	.DVKFGAD..A	NT
M. leprae	65- 84	YEKIGAELVKEVAKKTDDVA	+
M. tuberculosis	65- 84	....	+
Human	90-109	.KN...K..QD..NN.NEE.	-
M. leprae	85-109	GDGTTTATVLAQALVKEGLRNVAAG	+
M. tuberculosis	85-109	...............R.........	NT
Human	110-134	...........RSIAK..FEKISK.	NT
M. leprae	116-137	KRGIEKAVDKVTETLLKDAKEV	+
M. tuberculosis	116-137	........E........G....	NT
Human	141-162	R..VML..DA.IAE.K.QS.P.	-
M. leprae	153-171	DQSIGDLIAEAMDKVGNEG	+
M. tuberculosis	153-171	....	+
Human	179-207	.KE..NI.SD..K...RK.	-
M. leprae	195-208	KGYISGYFVTDAER	+
M. tuberculosis	195-208	...........P..	NT
Human	221-234	R....P..INTSKG	+
M. leprae	211-225	AVLEEPYILLVSSKV	+
M. tuberculosis	211-225	....D..........	+
Human	237-251	CEFQDA.V..SEK.I	NT
M. leprae	231-245	LLPLLEKVIQAGKSL	-
M. tuberculosis	231-245	.........G...P.	+
Human	257-271	IV.A..IANAHR.P.	NT
M. leprae	390-412	RKHRIEDAVRNAKAAVEEGIVAG	+
M. tuberculosis	390-412	....	+
Human	417-439	K.D.VT..LNATR........L.	+
M. leprae	412-425	GGGVTLLQAAPALD	+
M. tuberculosis	412-425	...........T..	+
Human	439-452	...CA..RCI.A..	NT
M. leprae	418-427	LQAAPALDKL	+
M. tuberculosis	418-427	.....T..E.	-
Human	445-454	.RCI.A..S.	NT
M. leprae	439-448	IVKVALEAPL	+
M. tuberculosis	439-448	....	+
Human	467-476	.I.RT.KI.A	NT
M. leprae	451-465	IAFNSGMEPGVVAEK	+
M. tuberculosis	451-465	......L........	+
Human	479-493	..K.A.V.GSLIV..	NT

+, positive; -, negative; NT, not tested.

T cells might be implicated in immunopathology. In contrast to the antibody data discussed in Chapter 9 in the case of T cell epitopes the sequence seems to predict specificity or, in this case, cross-reactivity. The two epitopes that are cross-reactive have at least eight identical amino acids or conservative substitutions, whereas the three epitopes that do not show T cell cross-reactivity with self have two or less identical amino acids or conservative substitutions. In fact Stefan Kaufmann and his collegues in Ulm have synthesized four mycobacterial 65 kDa epitopes with such a homology to self and shown that cytotoxic T cells reactive with these peptides do cross-react with self [8]. So these T cells reactive with autologous 65 kDa epitopes have not been deleted. This allows one to define the molecular basis for this kind of T cell mimicry, but it also implies that potentially dangerous T cells (think of adjuvant arthritis) need to be regulated, which will be the subject of the rest of this chapter.

HLA Class II Ir Genes and Immunopathology: Lessons From Leprosy

It was already mentioned in Chapter 8 that we have been using leprosy as a model for T cell mediated immunopathology and regulation by HLA-class II immune response genes [19]. Before discussing some recent data on regulation of T cell reactivity to the 65 kDa hsp in relation to this disease I will give some background information on the immunology and immunogenetics of leprosy.

Leprosy was endemic in Europe in the Middle Ages and in Norway even in the nineteenth century, which enabled Hansen to discover the causative microorganism, namely *Mycobacterium leprae* (*M. leprae*). Nowadays leprosy is a typical Third World problem with an estimated fifteen million cases. *M. leprae* is by itself virtually non-toxic and for most individuals a harmless parasite. In some individuals, however, disease develops and the majority of the symptoms of leprosy are attributable to the immune response of the host to *M. leprae*. If exposure to *M. leprae* does not result in T or B cell reactivity towards the microorganism, no immunity will develop but also no immunopathology. In the presence of AIDS this is an increasingly less theoretical possibility. The majority of individuals exposed and/or infected with *M. leprae* mount an adequate T cell response which results in immunity and no leprosy. However in some cases this T cell reactivity that leads to immunity also causes immunopathology, in particular nerve damage as seen in tuberculoid leprosy and so-called reactions. Finally, in some cases there is no T cell reactivity, and thus no immunity whereas the antibodies that are being formed in these patients affected with what we call lepromatous leprosy can result in immune complex mediated immunopathology.

Before Armauer Hansen discovered *M. leprae*, leprosy was generally considered to be a hereditary disease. For instance Simpson wrote in 1841: "few facts in the history of (tubercular) leprosy seem to be more universably admitted by all writers on the disease, both ancient and modern, than the transmission of the predisposition to it from parents to offspring" (quoted from [20]). After his discovery of *M. leprae*, Hansen argued that "a bacillary disease cannot be inherited" [21], and this simple concept has been popular in the first half of this century.

In the course of this century our knowledge of genetics and immunology has increased dramatically. The application of this knowledge to leprosy has provided evidence that the susceptibility to or expression of the disease are controlled by genetic host factors [19]. Notably the immunopathology resulting from the immune response to *M. leprae* is controlled by HLA genes, whereas susceptibility to the infection is not linked to HLA. Susceptibility to either tuberculoid or lepromatous leprosy is controlled by HLA genes and associated with certain

HLA-class II alleles, tuberculoid leprosy, for instance, with HLA-DR3 and lepromatous leprosy with DQ1 (19].

An important discovery, which may offer a clue to the association of HLA with leprosy type, has been the demonstration that HLA class I and II molecules are the products of so-called Ir genes. Ir genes code for genetically controlled differences in immune reactivity between individuals [22]. It has been shown that the products of particular MHC class I and II alleles may present certain antigens more efficiently than the products of other class I or II alleles and thus lead to differences in immune reactivity. Therefore it is easy to envisage that the extreme polymorphism of the HLA system may have evolved to a great extent by the vast selective pressure of virulent infectious agents and, in fact, we have presented evidence for that in the Introduction to this book. However, in the absence of these selective forces, we may be confronted with the disadvantages of such a selection. This may be true for both immunopathology related to infection with such a strikingly non-virulent parasite like *M. leprae*, as well as the large number of autoimmune diseases known to be associated with HLA alleles, notably class II alleles [23].

There is increasing evidence for HLA class II Ir gene control of the immune response to *M. leprae* [19]. How such HLA class II Ir gene control might explain the associations between particular HLA class II alleles and leprosy type is illustrated in Fig. 14.2. *M. leprae* reactive T cells are responsible both for acquired protective immunity to the bacillus and for the immunopathology observed in leprosy. Our Ir gene hypothesis predicts that different antigenic epitopes of *M. leprae* are presented by the products of different HLA class II alleles to functionally different T cell subsets causing either protective immunity and/or immunopathology. This hypothesis, which is currently being tested, would not only explain the association between particular HLA class II alleles and leprosy type and thus offer more insight into the immunopathogenesis of leprosy, but also

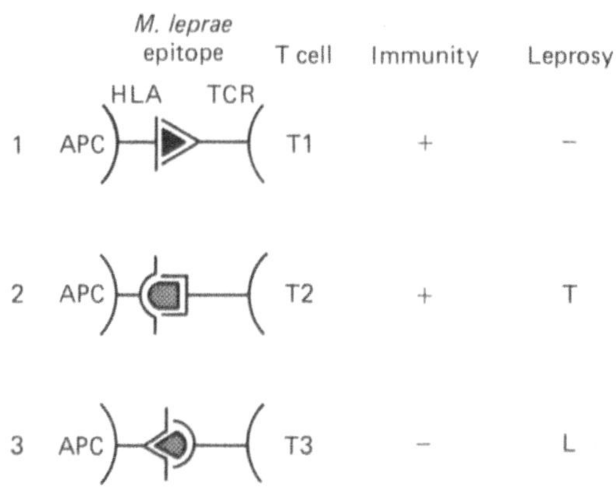

Fig. 14.2. HLA class II Ir genes and leprosy type. Antigen-presenting cells (APC) of three individuals differing for HLA class II present different *M. leprae* epitopes to T cell receptors (TCR) of functionally different T cells: T1 is a helper T cell that confers protective immunity, the same is true for T2 which, however, also causes immunopathology (delayed-type hypersensitivity) seen in tuberculoid leprosy, and T3 is a suppressor T cell responsible for the *M. leprae*-specific non-responsiveness seen in lepromatous leprosy.

has obvious implications for the development of a subunit vaccine against leprosy. Such a vaccine should:

1. Contain protective epitopes, that is epitopes recognized by T cells providing protective immunity.
2. Not contain disease inducing epitopes, that is epitopes recognized by T cells causing immunopathology (e.g. DTH T cells causing tuberculoid and suppressor T cells causing lepromatous leprosy).
3. Induce a protective immune response in all individuals.

In all these instances HLA class II Ir gene control may have to be taken into consideration.

HLA Class II Ir Genes and Regulation of T Cell Reactivity to the 65 kDa Heat Shock Protein

Because the 65 kDa hsp is a major antigen of *M. leprae* [24], we looked at the 65 kDa response of a large number of T cell lines from both patients and healthy controls reactive with mycobacteria [17]. The data are summarized in Fig. 14.3. The 65 kDa protein was presented via antigen presenting cells which shared only one HLA-DR allele with the T cell lines. The number of combinations are shown in the second row and the number of combinations that showed positive response in the third row. Quite unexpectedly, only HLA-DR1 and -DR3 restricted anti-mycobacterial T cell lines were reactive with the 65 kDa hsp. So, in other words, those class II genes control the response to the whole 65 kDa protein. We know in several cases (for instance for DR2 and DR5) that T cell clones reactive with 65 kDa hsp epitopes are present among the bulk T cell lines that do not respond, so apparently these T cells are downregulated. The second conclusion is that the N terminal epitope (aa2–12) is immunodominant for DR3 restricted T cells, which is also true for DR3-heterozygotes. We wondered whether the sequence would give a clue why this epitope might be so immunodominant, and in this case the answer is "yes".

As can be seen from Table 14.1, the aminoterminal (aa3–13) immunodominant DR3 restricted epitope on the mycobacterial hsp shows quite a substantial discrepancy with the human sequence. Notably, four amino acids in a row (KFGA) show non-conservative substitutions and these are in the core of the epitope. We know from a detailed analysis of T cell reactivity to this epitope that the core seven or eight amino acids are crucial for T cell epitopes. In seven epitopes that were not immunodominant we saw that at most two non-conservative substitutions were present. So perhaps tuberculoid leprosy (associated with DR3) might ensue because the response to non-self in this case is not so well regulated.

However DR3 is not associated with tuberculosis, whereas the DR3 restricted immunodominant epitope is shared between the 65 kDa hsp of *M. leprae* and *M. tuberculosis*. So, this DR gene phenomenon alone cannot be the whole explanation. Therefore we looked at DR3-restricted T cells restimulated with mycobacterial *M. leprae* in vitro and found to our surprise that, whereas T cells from

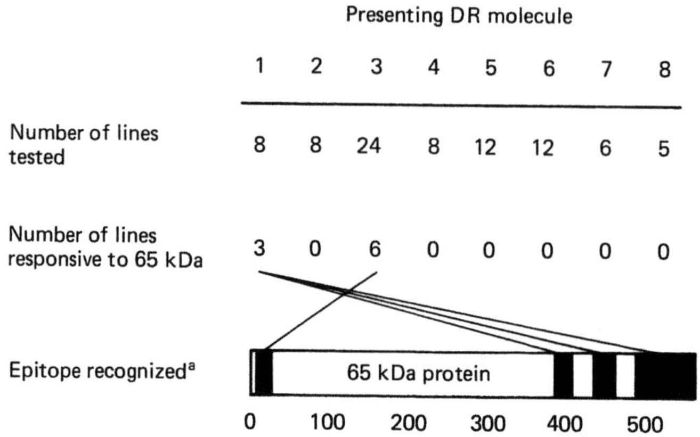

Fig. 14.3. HLA class II Ir gene control of T cell reactivity to the mycobacterial 65 kDa hsp. Fifty T cell lines from 48 naturally sensitized donors were raised by restimulating peripheral blood mononuclear cells (PBMNC) with *M. leprae* or PPD. All lines proliferated to *M. leprae* and/or PPD. They were tested for proliferation to the *M. leprae* and BCG (= *M. tuberculosis*) 65 kDa hsp presented by PBMNC as antigen presenting cells (APC) from donors matched for one of the two DR/Dw alleles of the T cell line donor. In total 84 combinations of APC-T cell combinations were tested. See Fig. 14.1 for the exact location of the T cell epitopes recognized.

healthy controls restimulated with *M. leprae* presented via DR3 were more reactive to *M. leprae* compared to T cells from the same individuals restricted via another DR-allele, the opposite was true in tuberculoid leprosy patients [25]. In other words, T cells from tuberculoid leprosy patients restimulated in vitro with *M. leprae* showed a lower response when the antigen was presented via DR3 as compared to presentation via the other allele. The same was true for the response to PPD. So in DR3 positive tuberculoid leprosy patients but not in healthy controls there is an *M. leprae* induced DR3 restricted low responsiveness to mycobacteria. John Haanen and co-workers have shown that this is due to *M. leprae* induced suppressor T cells, suppressing the response to 65 kDa heat shock protein. These cells are present in patients but not in healthy contacts (Haanen et al., unpublished work). The mechanism of this suppression is currently under study, but because these suppressor cells are thus far only found in patients, we conclude for the moment that the DR3 restricted response to the immunodominant epitope (aa3–13) of the 65 kDa hsp exerted by these healthy contacts may protect them from disease, in this case tuberculoid leprosy; in other words, that the T cell response to the 65 kDa hsp may at least in this situation be beneficial.

Acknowledgements. The studies reviewed in this chapter were performed in collaboration with Drs Rick Young (Whitehead Institute, Cambridge, USA), Jelle Thole and Jan van Embden (RIVM, Bilthoven, the Netherlands), Dave Anderson and Tom Buchanan (University of Washington and NeoRx, Seattle, USA).

Parts of "HLA class II Ir genes and immunopathology: lessons from leprosy" will also appear in Italian, in *Manuale di Leprologia* (OCSI, Bologna, 1990).
Financial support was obtained from the Immunology of Leprosy (IMMLEP) and Tuberculosis (IMMTUB) component of the UNDP/World Bank/WHO special Program for Research and Training in Tropical Diseases, the Netherlands Leprosy Relief Association (NSL) and the Dutch National Organisation for Scientific Research (NWO).

References

1. Lindquist S, Craig EA (1988) The heat shock proteins. Am Rev Genet 22:631–637
2. Ellis J (1987) Proteins as molecular chaperones. Nature 328:378–379
3. Flynn GL, Chappell TC, Roth JE (1989) Peptide binding and release by proteins implicated as catalysts of protein assembly. Science 245:385–390
4. Lakey EK, Margoliash E, Pierce SK (1987) Identification of a peptide binding protein that plays a role in antigen presentation. Proc Natl Acad Sci 84:1659–1663
5. Sargent CA, Dunham I, Trowsdale J, Campbell RD (1989) Human major histocompatibility complex contains genes for the major heat shock protein HSP 70. Proc Natl Acad Sci USA 86:1968–1972
6. Jindal S, Dundain AK, Harley CB, Singh B, Gupta RS (1989) Primary structure of a human mitochondrial protein homologous to bacterial and plant chaperonius and to 65kD mycobacterial antigen. Molec Cell Biol 9:2279–2283
7. Lamb, JR, Bal V, Mendez-Samperio P, Meklert A, So A, Rothbard J, Jindal S, Young RA, Young DB (1989) Stress proteins may provide a link between the immune response to infection and autoimmunity. Int Immunol 1:191–196
8. Kaufmann SHE, Schoel B, Wand-Württenberger A, Steinhoff U, Munk ME, Koga T (1990) T cells, stress proteins and pathogenesis of mycobacterial infections. Curr Top Microbiol Immunol 155:125–141
9. Young RA, Elliott TJ (1989) Stress proteins, infection and immune surveillance. Cell 59:5–8
10. Young D, Lathigra R, Hendrix R, Sweetser D, Young RA (1988) Stress proteins are immune targets in leprosy and tuberculosis. Proc Natl Acad Sci USA 85:4267–4270
11. Raulet DH (1989) Antigens for $\tau\delta$ T cells. Nature 339:342–343
12. Marrack P, Kappler J (1988) The T cell repertoire for antigen and MHC. IT 9:308–315
13. Bonneville M, Ishida I, Itohara S, Verbeek S, Berus A Kanagawa O, Haas W, Tonegawa S (1990) Self-tolerance to transgenic $\tau\delta$ T cells by intrathymic inactivation. Nature 344:163–165
14. Thole JER, van Schooten WCA, Keulen WJ, Hermans PWM, Janson AAM, de Vries RRP, Kolk AHJ, van Embden JDA (1988) Use of recombinant antigens expressed in *E. coli* K-12 to map B cell and T cell epitopes on the immunodominant 65kD protein of *Mycobacterium bovis* BCG. Infect Immunol 56:1633–1640
15. Lamb JR, Ivany J, Rees ADM, Rothbard JB, Howland K, Young RA, Young DB (1987) Mapping of T cell epitopes using recombinant antigens and synthetic peptides. EMBO J 6(5):1245–1249
16. Oftung F, Mustafa AF, Shuinick TM, Houghten RA, Kualheim G, Degre M, Lundin KEA, Godal T (1988) Epitopes of the mycobacterium tuberculosis 65kD protein antigen as recognized by human T cells. J Immunol 141:2749–2754
17. van Schooten WCA, Elferink DG, van Embden JDA, Anderson DC, de Vries RRP (1989) DR3-restricted T cells from different HLA-DR3 positive individuals recognize the same peptide (amino acids 2–12) of the mycobacterial 65kDa heat shock protein. Eur J Immunol 19:2075–2079
18. Dudani AK, Gupta RS (1989) Immunological characterisation of a human homolog mycobacterial antigen. Infect Immunol 57:2786–2793
19. de Vries RRP, Ottenhoff THM, van Schooten WCA (1988) In: Human leucocyte antigens (HLA) and mycobacterial disease. Springer, Seminar on Immunopathology, vol 10, pp 305–318
20. Fine PEM (1981) Immunogenetics of susceptibility to leprosy, tuberculosis and leishmaniasis. An epidemiological perspective. Int J Lepr 49:437–454

21. Hansen GA (1986) Heridity of leprosy. Arch Dermatol Syph 110:225–232
22. Benacerraf B (1981) Role of MHC products in immune regulation. Science 212:1229–1238
23. de Vries RRP, van Rood JJ (1988) HLA and autoimmunity. In: Cohen IR (ed) Perspectives on autoimmunity. CRC Press, Boca Raton, Florida, pp 1–9
24. Watson JD (1989) Leprosy: understanding protective immunity. IT 10:218–221
25. Ottenhoff THM, de Vries RRP (1987) HLA class II immune response and suppression genes in leprosy. Int J Lepr 55: 521–534

Subject Index

GPSR Compliance

The European Union's (EU) General Product Safety Regulation (GPSR) is a set of rules that requires consumer products to be safe and our obligations to ensure this.

If you have any concerns about our products, you can contact us on ProductSafety@springernature.com

In case Publisher is established outside the EU, the EU authorized representative is:

Springer Nature Customer Service Center GmbH
Europaplatz 3
69115 Heidelberg, Germany

Batch number: 10620881

Printed by Printforce, the Netherlands